ORANGUTANS

WIZARDS OF THE RAIN FOREST

Anne E. Russon

ROBERT HALE · LONDON

© Anne E. Russson 1999
First published in Great Britian 1999
By arrangement with Key Porter Books Limited, Canada

ISBN 0 7090 6615 5

Robert Hale Limited
Clerkenwell House
Clerkenwell Green
London EC1R 0HT

10 9 8 7 6 5 4 3 2 1

Electronic formatting: Jean Lightfoot Peters
Design: Peter Maher

Printed and bound in Italy

ACKNOWLEDGMENTS

My work with orangutans in Indonesia would not have been possible without the support and sponsorship of many individuals and agencies—first and foremost, Indonesia's Forestry Department with its Forest Protection and Nature Conservation Agency (PHPA) and its Institute of Sciences (LIPI), Dr. B. Galdikas and the Orangutan Research and Conservation Project, Dr. W. Smits and the Wanariset Orangutan Reintroduction Project (ORP), the International MOF-Tropenbos Kalimantan Project, the Biology Faculty of the Universitas Nasional (UNAS) in Jakarta, and funding support in Canada from Glendon College of York University and the Natural Sciences and Engineering Research Council.

My time in Tanjung Puting was greatly enriched by Mr. Ralph Arbus, Jane Fitchen, Valerie Gobeil, Andrea Gorzitze, Charlotte Grimm, many fine Earthwatch volunteers, ORCP staff (especially Pak Akyar and Pak Ucing), and Pak Baso of Kamai. At Wanariset, I am deeply grateful for the support I have enjoyed from everyone involved with the orangutan project, the Tropenbos project, and the Ministry of Forest. Thanks especially to Pak Boen, Pak Mulyana Omon, Pak Adi Susilo, M. de Kam, K. Warren, H. Peters, G. Fredrikkson, F. den Haas, P. Kessler, W. Tolkemp, R. de Kock, herbarium staff, A. Boestani, R. Siregar, and all of the ORP veterinary and clinic staff. Most of all, thanks to ORP technicians in Sungai Wain Forest, especially Dolin, Pak Misri, and Adriansyah who acted as my forest research assistants and without whose guidance, support, and companionship, I would have been lost—literally, more than once. Thanks to the Balikpapan Orangutan Society for its support, especially C. Luckett, M. Sowards, S. Randolph, and J. Cuthbertson, and to the people of Sungai Wain village for their generous hospitality, especially Pak Menaro and his family. Many people in Indonesia have helped create a second home for me—especially M. Polard, the Timpany family, the Tolkemp family, and the families of the town of Samboja. I am equally grateful for the support of friends and colleagues in Canada, especially C. Russell, J. Ankenman, E. Meyfarth, C. O'Connell, L. Spalding, P. Vasey, and the Bayview & Wilson Japan Camera, who helped me find the best in my photographs.

Lastly, and most importantly, none of this could have come to pass without the orangutans' willingness to allow me, in whatever small way, into their world. Getting to know orangutans like Princess, Unyuk, Supinah, Pegi, Daidai, Romeo, Enggong, Paul, and even Charlie has to count as among the greatest honors in my life.

CONTENTS

To my brother Bill, who first led me into the land of O's.

BORNEO AND SOUTH-EAST ASIA

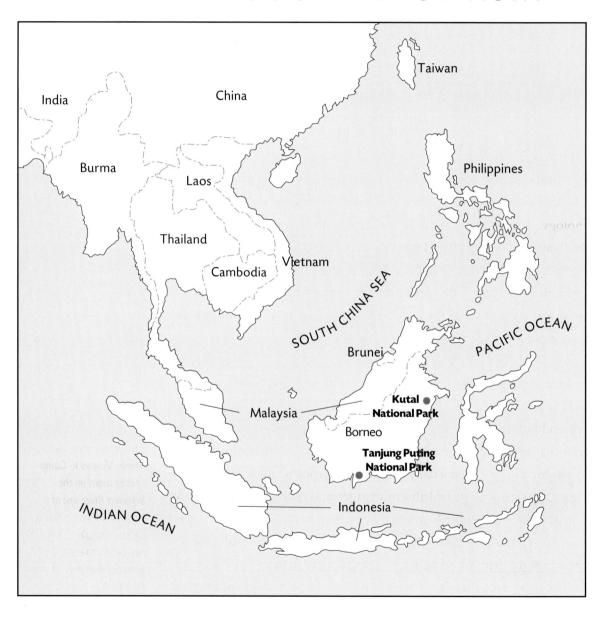

Taiwan

India

China

Burma

Laos

Philippines

Thailand

Cambodia

Vietnam

SOUTH CHINA SEA

PACIFIC OCEAN

Brunei

Kutal
National Park

Malaysia

Borneo

Tanjung Puting
National Park

INDIAN OCEAN

Indonesia

INTRODUCTION

It wasn't a dream. I really was on the roof of a weathered river house-boat, a *kelotok*, on a long trek upriver into the rainforests of Borneo, half a world away from my "real" life as a middle-aged Canadian psychology professor. The last stages of the journey that brought me here had been harrowing: a language I didn't understand, sweltering offices, interrogation by stern-faced uniformed authorities, and an alarming flight in a decrepit little plane, with cockroaches and an unshuttable bathroom door. At least my brother Bill was there, the only comforting presence in the vicinity.

We were on our way up the Sekonyer River to an orangutan research station, Camp Leakey, in Tanjung Puting National Park, Central Indonesian Borneo. We'd passed huge wooden sailing ships of another era, and an avenue of bowing nipa palms, before entering the narrow river that snaked into the interior. It didn't just *feel* otherworldly to me—it was. My usual life was a daily round of classrooms and academic arguments in a setting of snowdrifts, mortgages, and traffic jams. On the roof of that kelotok, I knew I'd never have gotten as far as the plane ticket without Bill's help. He was working in Sumatra at the time with CUSO, a Canadian aid agency, and he'd agreed to come with me.

The trip was to take four to six hours, making it feel like an incursion

Overleaf:
Rivers are the arteries of Borneo, and the only way into the forests where orangutans make their homes. Visitors to Camp Leakey travel on the Sekonyer River, one of Borneo's acidic blackwater rivers, its limpid, mirror-like surface tea-colored from the peat swamps that feed it.

deep into an unknown tropical jungle. In fact it wasn't far, something like twenty or thirty kilometers. The boats were just slow. But aside from the odd dugout lingering, languid, under the shade of the nipas, a few tipsy dwellings perched on stilts where river meets forest, and another kelotok moving logs, signs of human habitation were few. And they grew scarcer as we penetrated farther into the forest. To a newcomer like me, it all tolled *heart of darkness*, from the sinister stillness of the river to every strangling liana, brooding swamp, and slithering crocodile along the way.

I had once worked with chimpanzees, which had induced a deep interest in great apes and an equally deep distaste for laboratories. I wanted to study the great apes, but in free-living conditions, not in captivity. What intrigued me was their minds. I was interested in the age-old issue of whether they can reason, and if so, how well. The orangutans of Borneo were to be my first stop.

Large sailing ships sail in and out of the harbors of Bornean coastal towns. Here, they carry cargo from the port of Kumai to the other islands of the South Seas.

Let there always be orang-utans.
—Terry Maple, *Orang-utan Behaviour*

The orang-utan which I saw walked always on two feet, even when carrying things of considerable weight. His air was melancholy, his gait grave, his movements measured, and very different from those of other apes…signs alone were sufficient to make the orang-utan act; but the baboon required a cudgel, and the other apes a whip, for none of them would obey without blows. I have seen this animal present his hand to conduct the people who came to see him, and walk as gravely along with them as if he formed a part of the company. I have seen him sit down at table, unfold his towel, wipe his lips, use a spoon or fork to carry the victuals to his month, pour his liquor into a glass, and make it touch that of the person who drank along with him…All these actions he performed without instigation than the signs, or verbal orders of his master, and often of his own accord.
—Count Buffon, *Natural History: General and Particular, vol. 8*

ARISTOTLE'S RUBICON

For Westerners, the accepted wisdom about nonhuman minds is that they operate only at concrete levels; that they are so bound to the senses and physical events that they can never soar into the realms of reason and thought at which humans excel. Other species are said to live in the here and now, pushed and pulled in complex ways by the myriad events around them, but unable to plan beyond the moment. Nonhuman minds are deemed to work so differently from human minds that they constitute a different *kind* of intelligence, "animal" intelligence, which operates by a distinct set of rules and processes.

The great apes—chimpanzees (*Pan troglodytes*), bonobos (*Pan paniscus*), gorillas (*Gorilla gorilla*), and orangutans (*Pongo pygmaeus*)—have been considered to have the "animal" kind of intelligence. But great apes share many human attributes. It is because they are so like us, after all, that we send chimpanzees into space and conscript them into AIDS research. It should not be surprising, then, if their similarities include intellectual ones. By the time I headed for Borneo, researchers had amassed anecdotal and scientific evidence of great apes' powerful intelligence: they manufacture and use tools, solve problems by insight, and can acquire basic language skills. But much of that evidence was dismissed by hard-nosed scientists as mere clever tricks masquerading as something more sophisticated, and not sufficient to overturn the established assumption that they cannot reason.

A little questioning turns up good reasons to expect that great apes *would* have minds like our own. They are our closest living relatives. We share a common ancestor with all the great apes as recently as 12 to 15 million years ago, and with chimpanzees a mere 6 to 8 million years ago—mere ticks of the evolutionary clock. If any other species can reason, great apes are among the most likely candidates. Probing into history also exposes considerable evidence that our established wisdom on nonhuman minds could well be wrong about great apes. In fact, the scientific credentials of that "wisdom" turn out to be extremely shaky. In addition, great apes have been afflicted by the square-peg-in-a-round-hole syndrome. We have been more devoted to fitting them to our preconceived views than to fitting our views to the evidence before our eyes.

Gifts from the Greeks

The seeds of our views on great apes are commonly dated to Greek antiquity. It was Aristotle, in the third century BCE, who introduced ideas about the minds of animals versus humans. Aristotle assembled knowledge about animals from the four corners of the world to become the greatest thinker of his times on the nature of living things. So wide was his reach that he even wrote about apes, although Greece itself has none. The ancient Greeks had contacts with the parts of Africa and India where apes lived, so they knew of them, and had noted their similarity to humans. Aristotle had Alexander the Great, whom he had tutored, send information from Asia. Aristotle's studies led him to conclude that animals and humans were very similar in their mental processes, but only up to the Rubicon of reason. From what Aristotle could see, reason was the province of humans alone.

It makes a neat theory; it recognizes that humans share much with other animals, but marks a clear boundary between them. However, we now know that the information the theory was based on was far from complete. The whole world, as Aristotle knew it, wasn't very large. It didn't extend to the equatorial ranges of great apes—the African domains of gorillas, chimpanzees, and bonobos, and the Southeast Asian rainforests of orangutans. The apes Aristotle knew probably weren't great apes at all, but tail-less monkeys, Gibraltar's Barbary macaques. So Aristotle was ignorant of great apes when he defined the Rubicon separating human from other minds. We now know that great apes are humans' closest living relatives, and thus the key species to consider in determining similarities and differences between humans and other species. They are also the species most likely to bridge human–nonhuman boundaries. Theories of intelligence that don't take the great apes into account, Aristotle's theory included, are simply inadequate, because they are missing a crucial piece of the evidence.

The ancient Greeks also believed in near-human "monstrous races": Troglodytes, Pygmies, and Satyrs among them. Troglodytes were cave dwellers. Pygmies were a race of little people one cubit tall, the distance from the elbow to the tip of the middle finger. Satyrs were an evil four-legged race of forest dwellers, with human upper bodies, goat legs, and hair all over their bodies. They were lascivious, shameless, and wantonly inclined. Pan, their leader, was a great personage derived from the Greek god of flocks, shepherds, and woods. A playful lecher and a chaser of nymphs, he had a shaggy beard, and horns, and he walked erect with a staff. When Westerners saw real great apes, centuries later, they figured they had finally found the originals of the monstrous races.

Baggage from the past

At each step of the march through time, Western savants dredged up these flawed views from the ancient Greeks, accepted them on authority, added their own benediction, and forwarded them as established truth.

Pliny, an important Roman natural historian, revived the old tales of monstrous races. Early Christianity adopted the bits of Aristotle it liked (especially the part about only humans having reason) and dropped the rest, including Aristotle's appreciation of the similarities between human and nonhuman beings. It also exaggerated Aristotle's ideas, claiming an unbreachable boundary between humans and nonhumans, apparently as a tactic to distinguish Christian views from pagan ones.

Medieval attitudes blurred the distinction as thinkers again emphasized the similarities between humans and other species. They singled out nonhuman primates for their human-like intelligence—especially their imitative and problem-solving abilities—and placed them in a special category bridging the divide between humans and other species. At the same time, like the ancients they twisted what they saw to fit their preconceived view of things, The Great Chain of Being. This was another Aristotelian idea, the theory that all beings in the universe are ordered in their proper station, unchanging and unchangeable, from lowest to highest. To medieval thinkers, primates' human-like abilities were foolish or grotesque attempts to copy humans, and evidence of a presumptuous desire to rise above their proper station. They interpreted any such behavior in religious terms, as a sign of the Devil.

Great apes on the western stage

The West finally came face to face with real great apes in the seventeenth century, when Europeans took to the seas and knowledge of far-off lands began to trickle home. The first credited report came early in the 1600s from Battell, an English privateer who had been imprisoned in Africa for years. He returned to tell of two dangerous monsters common there, *M'Geko* and *M'Pungu*—Engecko and Pongo to European ears, now known to be chimpanzees and gorillas. Then in 1658 Bontius, a Dutch physician, described great apes he had seen during his years practicing in Batavia (now Jakarta). Great apes even began to arrive in Europe. The first we know of went to a Dutch menagerie in 1630; the next, a little fellow from Angola, reached London in 1697. These early traveling apes didn't survive long. The little Angolan died within months, from an infection said to have been precipitated by his falling against a ship's cannon during his voyage.

It was anatomists who took the greatest interest in them: Tulp, a Dutch physician and the anatomist in Rembrandt's *The Anatomy Lesson*, and Tyson, England's foremost anatomist, had started to dispel ignorance with their anatomical work, but then entrenched it even more firmly by identifying the apes with the

The identity of Bontius's ape is still disputed. Some claim that the beings he observed were hairy human beings, but as Herman Rijksen points out, in Bontius's descriptions they sound very much like modern orangutans.

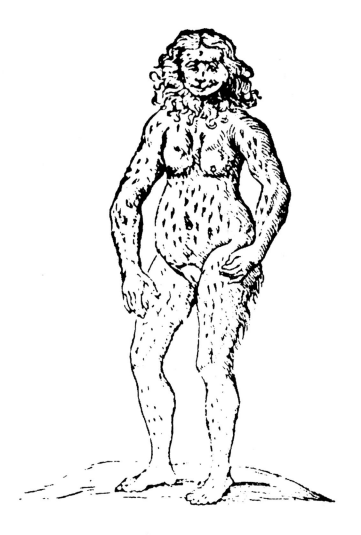

monstrous races. Tulp, in 1641, hailed his ape as Pliny's "Indian Satyr," a man of the woods, *Homo sylvestris*. Bontius and Tyson hopped on the bandwagon, claiming *their* apes were *Homo sylvestris* too. Tyson pronounced his to be *Simia troglodytes*, the mythical "Pygmie" of the ancients.

On top of mistaking living apes for mythical monsters, scholars were unclear about which ape was which. For over a century, the main mixup was between orangutans and chimpanzees, but only because they simply forgot Battell's gorilla. By the end of the eighteenth century, names had been invented and swapped back and forth so many times that naturalists had no language to discuss apes sensibly. Orangutans alone were lumbered with at least seventeen different scientific names. It was well into the nineteenth century before most of the essential distinctions were sorted out.

The identity of Tulp's ape is also disputed. Many believe it was a chimpanzee, because it was thought that it came from "Angola" in Africa. Herman Rijksen argues that it was an orangutan. It was thought to have been captured by a merchant from the East Indies, not Africa, so it may have come from "Angkola," Sumatra, where orangutans range today. Its features are more orangutan-like — a broad, smooth face, rotund stomach, perky little nose and missing toenails. It behaved like an orangutan too, walking upright and covering itself with a blanket when going to sleep.

The many names of orangutans

Simia Satyrus	Pithecus Brookei
Orangus outangus	Pithecus owenni
Pongo borneo	Pithecus curtus
Simia Agrias	Satyrus Knekias
Pongo wurmbii	Pithecus Wallichii
Pongo abelii	Pithecus sumatranus
Simia Morio	Pongo pygmaeus
Pithecus bicolor	Pithecus wallacei
Simia Gigantica	

Ape nature

An appreciation of great apes' nature, intelligence included, has been even slower in coming and more fantasy-ridden. Part of the problem must have been that the first great apes brought to Europe were commonly dead and pickled, well beyond thinking or doing anything. Another part of the problem must have been that Europeans looking at equatorial Africa and Asia faced a living world for which their beliefs did not prepare them, and which few wanted to accept. Even reputable scientists clung to the authority of old written words, in so doing succumbing to the human tendency to explain the strange in terms of the familiar. From the example of one ape, Tulp decided that great apes had satyric natures, and Tyson, also on the evidence of one individual (and his ape was dead), concluded that they stride along upright with walking sticks, like Pan. Tyson's stereotype stuck for over a century.

By the mid-eighteenth century, more orangutans and chimpanzees had survived the arduous sea voyage to Europe. Imagine the headlines: "Madame Pan charms London," "Satyrs in Paris." But even with living evidence before their eyes, Western sages twisted what they saw to fit the ideas of the day.

Some of their ideas were relatively benign. During the European Age of Enlightenment, late in the eighteenth century, thinkers like Rousseau thought great apes were the pristine men of the woods that humans were before our fall from divine grace and that their example could help us return to a more harmonious state. Orangutans were seen as the real link between humans and other animals and became *the* great apes to study for a short while.

Some of their other ideas were not so nice. With the nineteenth century came the Industrial Revolution. Work, commerce, and competition took over as central values. Nature was defined as frightening and of little value. Great apes were once again the wild men of the ancients, uncivilized brutes, ferocious and lewd, like Pan and his Satyrs. Orangutans were

LE JOKO. L'OURANG-OUTANG.

no longer noble near-humans but sluggish, lazy, and useless. The very names they had been assigned, orangutans as *Simia satyrus* and chimpanzees as *Homo troglodytes*, *Simia troglodytes*, or *Pan troglodytes*, no doubt fuelled the prejudice.

Scientific luminaries of the mid to late nineteenth century reinvented great apes again, this time to fit new notions about evolution. Lamarck and Darwin both thought great apes were the simian-like "ancestors" that had "fathered" modern humans. This set off the search for the "missing link," a man-ape or ape-man intermediate between human and ape. At the end of the century, that grail lured Eugene Dubois to sail for Southeast Asia in classic explorer style, in search of evidence of the man-ape. His first stop, in Sumatra, unearthed stunning fossil evidence of early orangutans but since he wasn't very interested in real great apes, he dumped that to pursue rumors of more human-like fossils in Java.

The twentieth century saw renewed attention to great ape minds from a newly powerful scientific community interested in exploring human minds from the perspective of evolution. Chimpanzees took premier status in this research as humans' closest living relatives; orangutans were backbenched. Chimpanzees are extraverted and dramatic, with quicksilver intellects. Slow-moving orangutans were soon stereotyped as the most obtuse, stupid, and boring of the apes. The best that might be said of that stereotype is that it may have spared orangutans the fate that befell many chimpanzees: unhappy lives as experimental subjects. Even that interest quickly faded as psychologists flocked to study rats, and anthropologists to study cultures and fossils.

Even the important eighteenth-century French naturalist Buffon fell for the "monstrous race" theory. He portrayed orangutans as walking upright with a staff.

Bornean views

There were other ways for early researchers to find out about great apes. The Dayak peoples of Borneo probably knew a great deal about orangutans, for instance, because they shared the forests with them. And Dayaks saw orangutans as powerful persons worthy of respect in themselves. Their accounts of meeting orangutans express awe and fear at orangutans' strength, invulnerability, and ferocity if attacked. Kenyah and Kayan Dayaks were careful not to offend them, and not to look straight at their faces or laugh at their antics. It was taboo among the Kayan for pregnant women to look at orangutans, and among the Sarawak Iban, to eat them. Head-hunting Dayaks collected orangutan heads as powerful symbols, perhaps for use in their animistic ceremonies.

The Dayaks saw in orangutans something very close to human. The Iban Dayak of Sarawak called them "forest cousins," and others described them as an inland race of people with whom fellowship and sexual union were possible. One legend tells of a child born to a male orangutan and a woman he had captured. The child was half orangutan and half human, its body divided horizontally. The very name given them, *orang utan*, means "person of the forest."

Some Dayak tales say orangutans *were* people, but they had fallen from grace for some blasphemy and were transformed into beasts as punishment. Another tells of their creation by two birdlike beings, the creators of all life on earth. One day the birdlike beings created man and woman, and they feasted late that night to celebrate their great accomplishment. The next day they tried to create more humans, but not feeling so well after the night's revels, forgot part of the recipe and turned out orangutans instead. As for orangutans not speaking, Dayaks of northern Borneo believed in orangutans' high intelligence and held that they *chose* not to, to avoid being made to work.

Some of these tales found their way to the West, but for the most part they fell on deaf ears.

Alfred Russell Wallace, co-discoverer of the principles of evolution in the mid-nineteenth century, saw orangutans with his own eyes on his voyages in the Malay archipelago. When he drew a picture of one, he chose to portray a male ferociously attacking natives. Few of his readers would have known that the natives probably attacked the orangutan first.

25

Even into the twentieth century, orangutans and other great apes were cast as the missing link between humans and other animals.

Looking past belief

Not all Western views of great apes were so hopelessly biased by scientific or religious preconceptions. Bontius's description of orangutan behavior is much like what we know today. Tulp's description of their head-covering, Rousseau's discussion of their ways of life, and Cuvier's assessment of their problem-solving abilities are also fair and accurate. A handful of early twentieth century scholars looked past old beliefs and found orangutans to be reflective, insightful beings. Yerkes, one of the most prominent great ape specialists of the early twentieth century, described them as nearly mechanical geniuses; Köhler spotted their capacity to solve problems by insight; and Furness and Haggerty observed that they could understand human language and learn by imitation. Furness ranked their minds within the human grade, perhaps only a step below the level of what were considered the most primitive human beings. But these thinkers' insights had little effect on mainstream twentieth-century thought about great apes, which agreed with Aristotle's conclusion that nonhuman species could approach the threshold of reason but could not cross it.

The long and the short of it is, we've been more attuned to what we *want* great apes to be than to what they are. The wherewithal to develop better understanding has been available, but ignored. For more than two hundred years, no one improved on Rousseau's description of orangutan nature, despite ample opportunity. Yerkes's findings on their mentality went ignored for almost fifty years. Old beliefs about apes are probably hard to overcome because of our resistance to any change in the boundary between human and nonhuman intelligence. That strict demarcation defines our ties with other species and largely determines how we treat them. Regardless of where we would like the boundary to lie, however, the evidence points to our having placed it incorrectly with respect to great apes. By the 1960s, at least some scientists figured we should take a proper look.

ORANGUTANS

101

I chose to study orangutans because most researchers in the field of great ape intelligence had ignored them, despite hints that beneath their sluggish exteriors they are exceptionally thoughtful. But if there wasn't much information on their intelligence, there was plenty on other facets of their adaptation; efforts to study wild orangutans began in the late 1950s. The job turns out to have its easy side. Orangutans spend their time slowly and serenely meandering through the treetops, just minding their own business. Unfortunately they do it in the almost inaccessible, swamp-logged, leech-laden tropical rainforests of Borneo and Sumatra. They also, on occasion, chase researchers out of their forests. It would take humans as obstinate as orangutans themselves to brave the discomfort of that environment to study them. Fortunately, there have been some. They now number in the dozens.

The great red Buddha

Mooch, an adult female, in direct light high in a kapok tree, shows her true colors.

Orangutans' signature feature has to be their red-orange color. Surprisingly, that is also one of the main reasons they are hard to study: it turns out that just finding wild orangutans is a monumental job.

Looking for a needle in a haystack is a kindergarten task in comparison. As a hint of how hard it can be, in the late 1950s George Shaller logged only 5.75 hours of direct observation, after months of work. Part of the difficulty is that orangutans come quietly, in ones and twos, way up in the canopy, half hidden in foliage, and on top of that, believe it or not, their orange color lets them vanish. Many researchers, myself now included, can attest to watching an orangutan disappear right before their eyes. The explanation lies in the way sunlight penetrates the forest. Because the rainforest is so deep and dense, most sunlight that filters through the canopy does so by bouncing off vegetation. Vegetation reflects green light—the color we see—but absorbs red and orange. By the time sunlight has bounced down through the canopy it has been robbed of its ability to register reds and oranges. In the midst of forest vegetation or on the forest floor, orangutans become large, dull brown lumps, just like the dead wood littered everywhere.

Beyond color, orangutans' most prominent feature is size. Like all great apes, orangutans are exceedingly large—in scientific terms, they come under the heading of megafauna. Largest by far are adult males, reaching 1.25 meters tall and over 100 kilograms in weight. Their strength is legendary, reputedly seven times as great as a man's. So much larger are males than other orangutans that the Iban Dayak people of north Borneo call them a separate species of orangutan, the biggest of three. The other

Approaching four years old, this youngster is dependent on his mother and incapable of surviving on his own; he is still considered an infant.

Tom, a young adolescent male already grown quite large, shows the stronger facial features of approaching maturity as he cavorts in the water.

two "species" are the middle-sized and the small (the Dayaks say they never get large). The middle-sized ones would be adult females and near-adult individuals. Adult females are close to a meter tall and usually about 35 to 40 kilograms, half the adult male weight. Little ones would of course be youngsters. They weigh only 1.5 to 2 kilograms at birth and remain under about 25 kilograms for six or seven years. In this phase they grow so slowly, they could seem permanently small. Their growth spurts at adolescence, but then they change so much as to be virtually unrecognizable.

Adult males also differ from other orangutans in having cheek pads, a great drooping throat pouch, and a distinctive "long call." Cheek pads are subcutaneous accumulations of fibrous tissue between eyes and ears that give their faces the look of giant

Above: The Great Red Buddha: an adult male.

Opposite page: An adult female carries her son, an "older" infant about three years old.

pieplates or blinkered horses. Peter Rodman and John Mitani have suggested they may help aim the male's long call so that it travels farther, and their visual impact—massive size—makes a very convincing threat. These features are likely important to orangutans' mating system, in which adult males compete for access to a precious group of females who can carry their offspring.

A subadult male shows why its strange body is so suitable for climbing trees. Subadult males are virtually adult size, but do not yet possess the cheek pads and throat pouch of full adulthood.

Orangutans, like all great apes, are designed for life in the trees suspended below branches, so their arms are longer than their legs. Orangutans are the only great apes to embrace this *modus operandi* close to full time, travelling as well as sleeping and eating in the trees, so they have exceptionally narrow, elongated arms and hook-like hands and feet for grasping branches. The other great apes, all from Africa, tend to travel on the ground by walking quadrupedally on their knuckles. When orangutans travel terrestrially, they walk either upright or quadrupedally on their fists.

Orangutans are exceptionally long-lived. The current estimate of their lifespan is thirty-five to forty years but researchers now know wild adults in their late forties *at least* and captives who have reached sixty. Orangutans have the longest immaturity of all the great apes, nine to twelve years. Their immaturity has three stages of roughly equal length—infant, juvenile, and adolescent—paced by birth, weaning, puberty, and sexual maturity. Females are deemed adult with the birth of their first infant, around fourteen to sixteen years, and males with the emergence of full adult regalia—cheek pads, throat pouch, and long call—as late as nineteen or twenty years. Males show an unusual two-stage maturation, with adulthood preceded by subadulthood, which usually starts when they are about ten years old and lasts until they are about fifteen, but sometimes continues until they are nineteen or even twenty.

Reproduction, correspondingly, is exceedingly slow. Orangutan females' estrus cycles average thirty (29–32) days, their pregnancies are just under nine months long (230–260 days), and they give birth to a single infant at a time. In all this they are very like human females, except for the fact that they bear an offspring only once every eight to ten years. This may contribute to the males' fierce competition for females. It also makes orangutans

A wild adult male rests in the cool of the dawn. Near the river rather than deep in the forest, he is unusually visible but in the early morning light, decidedly unorange.

extremely fragile as a species: they reproduce so slowly that losses take decades, even centuries, to replace.

Islands of the apes

Orangutans range today only on the islands of Borneo and Sumatra. They are not spread everywhere on the two islands. Their diet, primarily fruit, makes for a marked preference for lowland alluvial and peatswamp forest habitat close to waterways. The best orangutan habitat is coastal, because farther inland elevation rises and other forests predominate. Permanent orangutan populations are rarely found above 500 meters of elevation. Those ranging higher probably migrate to lower elevations for food for some periods of the year.

Orangutans' island distribution has led biologists to recognize two subspecies, Bornean and Sumatran. Sumatran orangutans tend to have red-orange hair, often whitish around the mouth and abdomen; Borneans tend to be darker, sometimes almost chocolate. Bornean adult males tend to have more robust bodies than Sumatrans as well as larger, differently shaped cheek pads, while Sumatran males have better beards. Sumatrans seem to be more arboreal and somewhat more social than Borneans, who have a bit of a reputation for turning into grumpy old men (and women).

The variability may be even greater. Borneo is the third largest island in the world, and is divided by high mountain ranges and wide rivers that, for orangutans, are as old and impassable as the straits dividing Borneo and Sumatra. Corresponding to Borneo's major geographical regions—east, northwest, and southwest—there may be three types of Bornean orangutan. It seems to be the easterners that tend to chocolate coloration. The southwesterners, in the region closest to Sumatra, look much like typical Sumatrans, although there may be several types of Sumatran orangutan too. Herman Rijksen detected a robust Sumatran type with lightish skin, orange-red hair, heavy build, and stubby hands and feet, often lacking nails on the thumbs and big toes, and a delicate type with darker, chocolate or maroon hair; dark, blackish skin; and delicately built, long, well-developed hands and feet with complete nails.

Being orangutan

Field researchers early spotted orangutans' priorities in life: eat, rest, travel, and occasionally socialize. The tempo of it all is slow, deliberate, and measured, and their acts are reflective, meticulous, and orderly. Orangutans virtually never stoop to the impulsive and reckless behavior of chimpanzees. They are among the mildest mannered primates—gentle, laissez-faire, and normally peaceable.

An adult female, Unyuk, with her infant son, Udik, who is about a year old. Some of the veteran rehabilitants at Camp Leakey had grown up and had their own youngsters.

The top of the list is food. As one expert, Biruté Galdikas, put it, the top three priorities in an orangutan's life are food, food, and food. They are frugivores, with 50 to 60 percent of their diet devoted to fruit. They consume a host of other foods as well, making a remarkably diverse menu of up to 400 different foods, including young leaves, sap, flowers, honey, shoots, stems, seeds, nuts, bamboo, fungus, pith, bark, soil, termites, ants, eggs, and invertebrates. Very rarely, they hunt for small mammals. Sri Suci Utami, an Indonesian researcher, once watched an adult Sumatran female search out, pursue, kill, and eat a slow loris (*Nycticebus coucang*), a small nocturnal primate. It takes vast quantities of food to fill their huge caloric needs, so they spend, on average, half their waking day just eating.

Orangutans rest a lot. It's easy to see how the story arose that they stayed mum to avoid work. They rise and retire with the sun; in the tropics, that makes 12 hours of rest and they often nap during the day as well. Every night, each orangutan constructs an intricately woven sleeping nest of leafy branches high in the trees. Orangutans are nomadic, so they usually build a fresh nest in a new spot every day—often near a good food tree that can provide the evening's dinner as well as the next day's breakfast. Even during the day they spend about 40 percent of their time resting, sometimes lounging in a quickly made day nest or draped along a comfortable branch or liana.

Orangutans travel over large areas of forest in search of food, each within a personal home range. Neighbors' home ranges often overlap, and adult females share their range with their pre-adolescent offspring. Bornean forests commonly support orangutans at densities of no more than one to three per square kilometer; even the richer Sumatran forests support at most six or seven orangutans per square kilometer. In Borneo, one adult female may need a home range as large as five to six square kilometers. Adult males' ranges are even larger, overlapping several female ranges.

Each day orangutans normally cover only a small portion of their

Whatever orangutans eat, they eat it in the most meticulous way.

A young orangutan eats a nangka. Fruit is the mainstay of the orangutan diet.

range, often traveling only as far and as fast as they need in order to eat. Sometimes they move as little as 200 to 300 meters and, on average, they whiz along at a third of a kilometer per hour—one tenth the speed of a chimpanzee. Adult males travel the fastest and the farthest daily, perhaps to satisfy their gargantuan food requirements, but perhaps also to control desirable females and intruder males.

Orangutan travel is almost entirely arboreal in Sumatra, where tigers still roam the forest floor. Large predators are fewer in Borneo, and there, the females normally travel in the trees, but adult males travel on the ground and climb to feed. Adult males likely travel on the ground because their weight precludes safe passage in the canopy. Like other apes, orangutans travel arboreally below branches, suspending their weight from their arms. Unlike other apes, they almost never jump or drop, and they rarely brachiate (swing by their arms from branch to branch). Instead they tree-sway or clamber their way cautiously through the canopy's tangle of branches and lianas. In tree-swaying, they ride slender pole-like trees as if they were vaulting poles, sailing in great arcs through the air from one huge tree to the next, or as if they were down escalators, letting their weight bow the tree right to the ground then stepping off neatly at the bottom. In clambering, they move by using any or all four limbs to grasp supports above, below, beside, and ahead. The look is almost of walking on air, or the oozing forward of a great red amoeba.

An adult female indulges in a few minutes of relaxation along a liana.

An orangutan settles down for the night in a sleeping nest that he (or she) has just built.

The greatest mystery surrounding orangutans is their largely solitary lifestyle. Solitude is unusual in primates, who are renowned for rich social lives. It is considered likely that orangutans have been forced into relative solitude because their enormous food needs mean that they can't afford to share. And no large predators except humans, and perhaps tigers in Sumatra, threaten adult orangutans, so they may not need groups for protection.

Mature males are the most solitary. They are so intolerant of one another that if they meet, they threaten each other by staring, inflating their throat pouches, long-calling, and shaking branches. If neither backs down the face-off may escalate to hand-to-hand combat, sometimes to the death. The only contact adult males appear to enjoy is a few weeks' consorting with a sexually receptive adult female. Such interludes are not frequent because females become sexually receptive only once every eight to ten years. Even then, the male abandons his female consort within days of her becoming pregnant.

Females are less solitary, even semi-social. They normally live with their offspring, making what is perhaps the only stable social unit in orangutan society. Immatures are downright gregarious. Infants stay in constant contact with their mother for the first year of their life, venture only meters away during their second, and travel and sleep with her until the approach of weaning at four to six years of age. Even as juveniles they remain semi-dependent and tend to range near their mothers. Youngsters may see little of older siblings because of the long interval between births, but juveniles establish friendships with same-sex age-mates. Youngsters start to move off on their own as they approach adolescence, males at around seven to eight years of age and females a bit later. Adolescent females commonly establish their own home range near their mother's. Adolescents too can establish long-term relationships, often with opposite-sex partners.

Orangutan sexuality shows yet more quirks. Unlike many primates, female orangutans are willing to mate throughout their menstrual cycle and show no obvious signs of estrus. When an adult female is inclined to mate, she exercises considerable choice. She may actively seek out the particular male she desires, listening for his long calls to help locate him; then she will entice him with a host of seductive tactics, and actively stimulate his arousal. As for undesirable males, especially subadults, females just as actively spurn their advances. Persistent rebuffs may account for the fact that subadults regularly resort to forced copulations, which have been likened to rape.

*An adult female and her son
clamber through the trees.*

Thinking on the wild side

It may not seem that the demands of orangutan life are very complicated—perhaps, you are thinking, they have no need of high intelligence. That impression could be one reason orangutan intelligence has not been prominent on the field research agenda. Another reason could be that the close observation needed to study intelligence is rarely possible with wild orangutans.

Most field researchers have nonetheless noticed signs of sophisticated thinking. Orangutans' tree-to-tree movement shows an exquisite understanding of how to manipulate multiple forces. Their travel routes suggest that they navigate by detailed "mental maps," which include not only the location of food sources but also the distances, routes, and obstacles between them. They have been observed to chart the shortest route to their destination from their current position. They use "traditional" travel routes, routes known and used by many individuals, suggesting some sort of culture. They seem to plan their travel routes in advance, establishing clear travel directions that take them to obvious goals. They may also detour from their current route to reach a known food tree, check its crop and, if they find it not ready, backtrack to their original route. Feeding may also require high intelligence because some orangutan foods, such as coconuts and spiny durian fruit, are very difficult to obtain. Orangutans have been seen to use very complex techniques to extract such foods despite the plants' apparently impregnable defenses. But this ties in with one of the great black marks against orangutan intellect: wild orangutans, unlike their chimpanzee cousins, rarely use tools to gain access to difficult foods.

Understandably, then, most scientists interested in studying orangutan minds have turned away from wild orangutans in favor of captives. Interest in nonhuman primates perked up enough in the late 1950s and 1960s to spawn a few studies of orangutan intellect, even though orangutans were sidelined once chimpanzees took center stage at the turn of the century.

In these studies, orangutans have repeatedly demonstrated the mechanical genius that had so impressed Yerkes, mastering tools of exceptional sophistication and solving complex physical puzzles. One mastered making stone flake tools, which he used for cutting. Another tackled a complex locks-and-boxes task. He was presented with five boxes, one of which held keys to the other four. Of the four locked boxes, one held food, one was empty, and two held keys—*one* of which would unlock the food box. To get the food, he had to select the right key from the first box—the key to the box with the key to the food box, unlock that box, take that key, then unlock the food box. He breezed through the test. Orangutans seem to solve some of these problems by insight, that is, by thinking, rather than by trial-and-error fiddling.

Orangutans also (again) showed capacities for language, lying, planning, and imitation. One famous orangutan, Chantek, did everything. For eight years, starting in 1978 when Chantek was nine months old, anthropologist Lyn Miles tutored him in the American Sign Language for the deaf. Great apes are incapable of speaking but have been apt pupils of sign language, a language of gestures. Lyn deliberately immersed Chantek in the world they talked about to give him the same advantages as a human child learning language. That world was Chattanooga, Tennessee. So Chantek went to fast food joints, ate hamburgers, chose his own ice cream flavor, had chores like cleaning his room, and got an allowance for doing them. He learned more than 150 signs, and he made and understood short sentences, rivalling the accomplishments of bonobos and chimpanzees. He lied about three times a week, mostly to get into the bathroom. He'd sign "dirty," meaning he needed the toilet, but once in the bathroom he just wanted to play and wouldn't use the toilet at all. He recognized himself in mirrors. He even liked to check out new "looks" in mirrors, for example, how he looked in sunglasses. He planned, signing "in milk raisin" to himself before heading into his trailer and asking a caregiver for "milk." And he imitated: he curled his eyelashes at the mirror with his caregiver's eyelash curler, just as she did, and he cooked things—like his cereal, which he put into a pot, then onto the stove, and his camera, which he put into a pot of water.

These studies put orangutan minds on a par with the minds of other great apes, maybe even beyond. Many of their accomplishments— insight, tool use, self-awareness, language, planning, imitation—are on the reasoning side of Aristotle's Rubicon. Some scholars discount these accomplishments as illusions, the artificial products of unsound research. They argue that great apes accomplish these feats only with extensive human support, so what is shown is the power of human culture, not the power of apes' minds. But you could just as easily argue that these findings *under*represent apes' abilities. Great apes' lives in captivity are more often marked by deprivation than by enrichment. And most findings come from experimenter-planned tests and command performances that are probably about as inspiring as having to play the piano for family guests. Great apes' abilities could, equally plausibly, be better than what the studies of captive subjects have shown.

Personal pickle

What I wanted was to study orangutan minds functioning at their peak and that put me in a bit of a pickle. Ideally, wild orangutans in their natural setting would be my best bet; only there would I be able to watch them coming to grips with the real intellectual problems they and their ancestors had been dealing with for millennia. However, their lifestyle makes the close-up, fly-on-the-wall view I needed almost impossible. Captives, on the other hand, are accessible but often emotionally and mentally troubled as a result of human treatment, and thus unlikely to function at their best. On top of that, there are few bridges between these two forms of intelligence—the first developed to handle the problems of rainforest life, the second to handle a rarefied set of human-defined situations.

What I needed was orangutans who could help me bridge this gap, by bridging it themselves. And I found them, although I didn't appreciate the richness of my find at first. The orangutans I had come to Camp Leakey to study were ex-captive orangutans under rehabilitation to free forest life, and thus belonged to a third, intermediate category. What I knew of them was they had been captured to be kept as pets or sold; there is a lucrative and illegal trade in live animals. Any captured orangutans that are located by the authorities are confiscated and turned over to special sanctuaries in Borneo and Sumatra for rehabilitation to free forest life. I thought these rehabilitants might have the qualities I sought: they would be familiar enough with humans to tolerate close observation, and knowledgeable enough about human customs to act in ways I could comprehend. They had also been released from the deprivations and frustrations of captivity so I expected to be able to observe involved, active, functioning individuals rather than bored, depressed, or abused ones. And in the sanctuary environment they would be free to develop their abilities in tune with their own interests and the world in which they were designed to live. These were my hopes, and as it turned out my expectations were correct.

Little had been written about rehabilitant orangutans. There has been a tendency to dismiss them as damaged goods, no longer "pure" wild orangutans but oddball nutcases, fit only for the odd anecdote or the education of visitors. One of the few serious reports I found came from Camp Leakey, the center in southcentral Borneo run by Biruté Galdikas. Galdikas is a well-known Canadian, a pioneer in the study of wild orangutans and the third of the three women "angels" that Louis Leakey sponsored to promote the study of free-living great apes. Like Jane Goodall's work with chimpanzees and Dian Fossey's with mountain gorillas, Galdikas's work with orangutans became a lifetime project. She had been studying wild orangutans since the early 1970s and rehabilitating ex-captives for almost as long. Her articles described rehabilitants as great tool users and suggested they learned by imitation, the ability that interested me most. And that was how I came to be on board a kelotok in August 1988, heading upriver into the rainforests of Borneo.

THE WORLD ACCORDING TO CAMP LEAKEY

The kelotok made slow progress, steadily moving past banks of rich vegetation. Late in the day, the river opened onto a sunny, marshy area and then, within minutes, we were at Camp Leakey. All we could see at first was a long, long ironwood dock that stretched across an open pan of riverside scrub and disappeared through a curtain of forest. We disembarked and started up the dock. It almost barred entry, forcing us under the glare of the equatorial sun, its loose planks clattering at every footfall. But past the dock and through the forest curtain, the camp was an unassuming haven, a few wooden buildings painted white with blue window frames, nestled in the woods. Our clattering alerted its residents, who came out to greet us. There were about twenty Dayak field staff, a team of foreign volunteers, a Canadian colleague of mine, Biruté Galdikas herself, and, eventually, a number of orangutans. We stayed only five days, just enough to taste the camp's flavor and discuss research possibilities.

Camp Leakey had always been more than a rehabilitation sanctuary. Its original purpose was wild orangutan research and it was buzzing with research activity when we arrived. Galdikas's thirty-odd staff members were all busy collecting data on wild orangutan behavior, monitoring the forest's life cycles, or running the camp. Individual researchers and

Kelotoks arriving at the Camp Leakey dock were often met by one of the camp's many rehabilitant orangutans. This time it's Tom, adolescent son of an ex-captive female named Tutut.

50

students were sometimes in residence with their own projects, which explained the presence of my colleague, Anne Zeller, an anthropologist from Waterloo. She was studying the long-tailed macaques that ranged around the camp. The volunteers were courtesy of Earthwatch, an American organization with an innovative system of supporting conservation-oriented research. It organizes teams of volunteers to assist the research, and helps fund the work to boot. Its teams arrived at Camp Leakey several times a year for two-week work parties. Not only did they assist with live research, they also went on educational treks into the forest and to a local gold mine, and attended daily lectures on orangutans and related topics such as local culture or threats to orangutan survival. Ad hoc activities like patrolling for poachers, helping care for infant ex-captives, or attending a Dayak wedding might round out the volunteers' schedule.

The camp had a sleepy feel about it during the day. Life moved into the forest then, orangutans foraging and humans researching. But almost everyone returned to camp late in the day and then it started to buzz. A motley crew of ex-captive orangutans ambled back late in the afternoon, for the provisions on the dock and to nest. Humans returned for much the same reasons: we too were provisioned, except our provisions were called meals and they were at the dining hall, not the dock. When our dinner gong sounded through camp, around eight p.m., human affairs sprang to life. Little files of people left their quarters and picked their way along the pitch-dark web of paths, flashlights bobbing, to assemble at the dining hall. There Galdikas presided over dinner and the nightly ritual of reporting work. She called on each team to report its day's activities, questioned their report, carefully recorded it in her log, and after reflection, she assigned the next day's tasks. Then music, story-telling, or a talk might while away a relaxing hour before lights went out.

I was there to size up the rehabilitants as subjects for my research, so my time went to making their acquaintance, looking at the rehabilitation program, and talking with Galdikas. The program focused on re-educating ex-captives to forest life, a slower and a more difficult task than one might expect. Re-education faces two hurdles: the orangutans had to acquire forest skills *and* unlearn counterproductive human habits; both could be a challenge. Instilling the will to turn from human to forest ways could take years, especially in ex-captives whose earliest years were spent with humans. Instilling forest expertise is a multi-year task too. Among other things, human "teachers" don't have the expertise themselves—neither knowledge about orangutan forest foods nor the physical capacity for life in the trees. There was also much to learn and the orangutans learned it the way they did everything: slowly.

But rehabilitation had been operating at Camp Leakey since the early 1970s, and it had a well-worked-out, if casual, system. Given the difficulty of teaching orangutan expertise, Camp Leakey offered the ex-captives the opportunity to learn in the form of an easygoing bridge between the their two worlds, forest and human, plus a fall-back cushion in case of hitches. Ex-captives could indulge in one, the other, or both.

Orphan ex-captives too young to wander on their own, those under about five or six years of age, were cared for in the camp nursery. They were kept in protective caging when they couldn't be supervised, sent out to forest school each day to practice climbing trees, making nests, and eating bark, and assisted with additional provisions. Replacement mothering was shared among the camp's Dayak nursery staff, who played teacher-protector, and other orphans in the nursery, who gave what emotional support they could. Occasionally, with a little encouragement and a lot of luck, an older ex-captive adopted an infant orphan. Adoptive "parents" were mostly adolescent females who were starry-eyed over adorable infants, or adult females who'd recently lost their own infant, although a subadult male apparently once made the offer.

Older ex-captives were simply free. Their lives in the rehabilitation program and around camp were up to them, carte blanche, notwithstanding a few house rules like no breaking into buildings. There were no barriers around Camp Leakey other than the few protective cages, so

In the Camp Leakey environment, humans and orangutans were all part of one world, without boundaries or bars. Tom and Nancy (the son and adopted daughter of Tutut, an adult female rehabilitant) join Ralph (from New Jersey) as he carries Doyok (a new arrival) down to play by the river.

ex-captives had free run of forest and camp, freedom of association with orangutans and humans, free provisions daily, and virtually any support they needed. The idea was that they would follow the normal orangutan pattern of gradually, with age and experience, exploring and ranging farther and farther away from their mother—even if "mother" was another hapless orphan or a hired human substituting for the real thing. When they felt ready for independence, they'd just mosey off of their own volition. The system worked, in a casual and partial sort of way, as long as you weren't insistent that ex-captives abandon human ways.

By 1988, the rehabilitant scene was complex. The rehabilitants had developed a rich community that spanned all ages and stages of rehabilitation. According to Galdikas, more than a hundred ex-captives had passed through her program in her twenty-five years of operation.

Many had long since vanished, but as many as thirty might still show up for provisions on any given day. Refugees from captivity were still arriving, mostly new charges for the nursery. When Bill and I arrived the nursery had four or five waifs in residence, from infants small enough to fit in your hand to juveniles ready to make the leap to independence. Depending on their individual personal history, they either tumbled around the forest playground, huddled in misery, or desperately clung to anyone who'd let them.

Tutut, a rehabilitant orangutan, and Yul, a rehabilitant gibbon, casually ate together on a daily basis.

Older juveniles beyond the need for protective care ambled about through the camp and nearby forest, wreaking various forms of orangutan havoc. Herbie, a male about four, thought it was cool to lurk in the bushes just off paths and tackle unwitting passersby for a hearty wrestling and biting session. Few of his human victims shared his amusement. Adolescent and even adult orangutans also ranged around camp, most of them veterans of rehabilitation from the early days. Several females had offspring, some two or even three, and a couple were expectant grandmothers. A whole community of unusual youngsters was developing—never captive themselves, raised by their orangutan mothers, but quite at ease with humans.

Even independent ex-captives and their offspring still had access to provisions, if they wanted them. In 1988, provisions were offered at the dock twice daily, with extra snacks available on request. Some older ex-captives had done as the camp's founders had hoped, and acquired enough forest expertise to move on

Siswi, an adolescent female, rides in the camp's "non-human primates only" swing.

to a self-sufficient life in the forest. They still might pass through camp from time to time, once every few months, but they mostly stayed for a few days' provisions then vanished again. They simply used camp as an orangutan-style fast food joint. Some, however, stuck to camp and made it their regular routine to head to the dock early in the morning for provisions, shuffle off on their daily rounds into the forest or camp, and meander back to the dock in time for afternoon provisions, then retire to a nearby nest.

Other forest creatures had also discovered the camp and its provisions. Wild orangutans, long-tailed macaques, bearded pigs, civet cats, barking deer, monitor lizards, and snakes periodically congregated there. The camp also provided sanctuary to other ex-captives, from agile gibbons and pigtail macaques to hornbills and an infant sunbear.

For me, the prospects looked good. Ex-captives were as amiable, unconcerned, and approachable as I could wish. I could stick my nose into their affairs without provoking so much as a twitch. Some were "bicultural," moving back and forth between camp and forest problems and between human and orangutan companions. Some tackled human problems on a regular basis, making it easy for me (a human) to understand the complexity of their thinking. Provisioning took the weight of foraging off their backs, leaving them extra time to indulge their curiosity and other intellectual pursuits. They lived in such a rich multi-species society that their social lives were very evident. I was beginning to think that—other than the fact that they moved at the rate of molasses in winter—they would be a good group to study.

I had enough of an inkling of these qualities to ask Galdikas for permission to return the next year to study imitation. But it was really my brother Bill who appreciated the camp's uniqueness. There I'd be, he'd say, walking down the path going about my business, I'd meet an orangutan coming the other way down the same path, going about her business, and we'd just pass each other as if nothing special had happened! What Bill saw was a world where everyone—humans and nonhumans alike—mixed together in gay disregard for boundaries of the species kind. What Bill saw, and I later came to appreciate, was that life in Camp Leakey had the look and feel of the bar scene in *Star Wars*.

Bridging two worlds

The next year, I took a close look at the world according to Camp Leakey. With the easygoing support that was available, each orangutan found his or her own individual path to free life. As for the expertise needed to survive in the forest, ex-captives approached the task with what you might call a "human accent."

FOOD. At the top of the rehabilitation priority list for ex-captives is learning to forage for forest food. Fortunately, orangutans are food addicts, so most ex-captives sooner or later cottoned on to the fact that the forest amounted to a twenty-four-hour-a-day free restaurant. Equally, however, ex-captives zeroed in on other food opportunities. The most obvious of these was the faithful daily provisioning. Provisions were intended for ex-captives not yet able to forage for themselves, but it appeared that to the orangutan way of thinking even an ex-captive who was skillful in forest life would be silly to pass up the chance of easy food. I met no silly orangutans. Some ex-captives, especially those that hung around the camp, also discovered stashes of food intended for humans and devoted at least as much energy to obtaining

Like humans, Tutut, an adult female, enjoyed a mug of coffee; here, she begs some from Ucing, one of Camp Leakey's longtime staff and a good friend of hers.

them as they did to forest foods. Many of the "human" foods were "better" than forest ones, richer in calories and ready to eat, so, once again, ex-captives who aimed for them were being smart. Some mastered picking locks, and could open the doors to places holding food whenever they liked. The better-mannered had the decency to close the door behind them when they left.

Some ex-captives had worked up compelling begging routines. Orangutans do beg food from one another in the wild, so this was, in a sense, normal; less normal was *how* they begged. Some had a taste for coffee, and of course there was no way they could get it on their own, so if they spotted a human with a mug of coffee, they'd find an empty mug and hold it beseechingly towards the coffee owner, or attempt to "buy" the coffee with a "payment" of leafy branches. Refusing to share on the grounds of "too hot" didn't foil the beggars, because they'd also figured out how to cool the coffee themselves—blow over its surface, sip it through a straw, pour in some cold liquid, or dip and sip small spoonfuls.

Sugarjito wandered into the camp's outdoor kitchen one day and came up with the idea of using the wok as a prefab nest.

Many ex-captives knew how to use human food utensils and some even seemed to prefer them. Given the option, they drank river water from mugs or bowls rather than sipping it directly, wild style, and ate messy foods with plates and spoons, as if to keep fingers clean. A few of the more enterprising even tried cooking. Galdikas recalled one who'd tried to make pancakes. She broke eggs, put them in a cup, added flour, then mixed the mess. The only thing she didn't do was actually try to heat them.

NEST AND REST. Ex-captives had no problem embracing the orangutan ideal of conserving energy and enjoying rest. When hunger wasn't too pressing, they spent much of their time lolling about taking it easy. For sleeping at night, they were expected to adopt the forest pattern of making nests in trees. Most did, but they added unusual quirks. Those

Hammocks aren't normally used in Borneo, but when Western visitors imported them, orangutans quickly discovered their benefits.

that liked to sleep near camp had long since used all the best nesting sites and denuded the trees of foliage. That meant that the normal orangutan system of building a new nest at a new site every night was no longer an option. So ex-captives took to re-using old nests and the camp area came to have a number of established nest sites. These sites operated rather like orangutan hotels—climb in and it's yours for the night. There were even double-decker nests for mothers with youngsters, a large sturdy one at mid-level with a little one just above.

Re-using old nests made a few new challenges. Leaves soon dried up, probably making them uncomfortable. So most ex-captives gave old nests a quick overhaul with a fresh bouquet of leaves, collected on the way up. Some even brought bags to carry their leaf collection. Others preferred linen. A frequent visitor, Mr. Ralph, once groused at staff for not returning a sheet he'd sent for washing. Staff said it had vanished, and that perhaps an orangutan took it. Mr. Ralph scoffed at that excuse until someone returned from the dock reporting having sighted said sheet, a hundred meters downstream, aflutter atop an orangutan nest. A worse problem with old nests was that other critters moved in after a few days. More than a few ex-captives settled down in an old nest only to leap out a minute later, bitten by stinging ants or startled by a colony of bats streaming out from below.

Another solution to the nest problem was to loosen up about sleeping sites. Some ex-captives took over the camp's observation tower at the river's edge. It had a winding stair-case, with landings at each turn, that led to a lookout pulpit at the top. Landings and look-out made good sleeping platforms so the tower became another orangutan hotel. The lookout was the penthouse, the lowest land-ing, a budget flat. Orangutans checking in above their rank risked being bumped down if a higher-rank companion arrived. A few inven-tive ex-captives concocted even more unusual solutions. One adult female, Supinah, some-times slept in a little house near the camp nursery, closing the door once inside. Another, Princess, occasionally slept on the doorstep of the bunkhouse porch. Last but not least, many ex-captives discovered hammocks. The more brazen simply climbed in with the hammock's human occupant. At that, most humans scrambled out, fast and fearful, leaving the orangutan in sole possession. But humans soon figured out that they could rid their hammock of unwanted orangutans by untying

Unyuk walking upright. Even wild orangutans are known for walking upright, but Unyuk, who grew up with humans, took it to extremes.

it so that it fell to the ground. Orangutans didn't know how to hang hammocks up on their own, so that finished the game and they left. Unfortunately, a few of the most ingenious ex-captives finally figured out how to hang hammocks up, so the game was back on.

TRAVEL. Ex-captives did take readily to the normal orangutan system of moving through the trees. But liking treetop travel doesn't necessarily mean being good at it. Some of the more earthbound ex-captives put on a pretty embarrassing show when they tried arboreal travel, breaking every second tree they climbed through. Overall, even with years of rehabilitation, ex-captives travelled on the ground more than wild orangutans do, walking upright even in the woods, and using trails to navigate.

Ex-captives also had some unusual travel strategies they must have picked up from humans at camp. Ex-captives regularly stole camp canoes for "pleasure rides" downstream. Humans coming upriver occasionally passed one of the ex-captives cruising down in a camp canoe. It wouldn't have been a big deal except the orangutans didn't bring the canoes back. They just hopped off at their destination and left the canoe floating on down. So humans tried to stop the pleasure-boating by tying canoes to the dock. Orangutans are knot experts, though, and ex-captives could untie every knot that staff could make, even triple and quadruple ones. Eventually staff resorted to sinking canoes. As of 1988 that had stopped the canoe junkets, or at least most of them. Other orangutans were apprehended trying to start motors on speedboats; fortunately, at least as far as I've heard, they haven't yet succeeded.

SOCIALITY AND SEXUALITY. If ex-captives invented their own rules for camp/forest life, they also set their own social and sexual ones. Orangutans are supposedly lovers of solitude, indifferent to the company of others. Around Camp Leakey, rehabilitants lived in a large community of kin, friends, foes, and occasional mates, all with well-defined social roles. Akmad and then Siswoyo reigned over the adult females.

There were so many orphaned infants in camp at one point that the easiest way to take them to the river for a bath was to wheel them down in the camp wheelbarrow.

Here Nancy, a rambunctious juvenile female, uses pilfered rice sacks to add variety to her dangling and spinning.

Siswi, Siswoyo's adolescent daughter, enjoyed high status via her mother. Supinah and Princess, also adult females, ranked low—probably because both were still quite young, and as orphans had little orangutan family social support. Perhaps it was partly for that reason that these two sought the company of humans. Humans treated them better than their fellow orangutans did.

Ex-captives' social lives weren't all status oriented. Many had personal friends with whom they'd travel, forage, and play in the forest. Davida, a juvenile-going-on-adolescent female, often spent days with a favored infant or juvenile. Even adult females teamed up as travelling mates; so did subadult males. Some of these friendly ties, I was told, dated from having lived together as young orphans in the camp nursery. Ex-captives might even travel in large groups, not just the typical twos and threes. One such travelling group grew into a carnival of thirteen orangutans all tumbling and playing together.

The ex-captives' society was not a closed club. It was semi-open to humans, in odd ways. Siswi, for instance, was never captive—she was born and lived free—but she grew up around camp and mingled easily with humans. She eventually took the role of camp hostess, meeting human visitors and escorting them on tours of camp and its environs. Supinah, too, treated humans as part of her circle—or herself as part of the human circle, depending on how you looked at it. She seemed to see herself as relatively high in rank among human females: she behaved with normal restraint around dominant women, like Galdikas, but nipped or even attacked lower-ranking ones, like cooks and volunteers.

Sexuality was another matter that took a unique turn in this unique society. Orangutans, like many species, tend to develop an unusual sense of who they are if they are raised outside their own species. Rather like Supinah, human-reared orangutans often saw themselves as more human than orangutan. When orangutans like this mature, their affections sometimes fix on their adoptive species. Always flexible, though, the rehabilitants did not allow any predilections they had for humans to interfere with their interest in one

Herbie loved to be carried. Here, his "mark" is an Earthwatch volunteer. Unfortunately, Herbie was at this point a 25-kilogram adolescent and he would refuse to get down. Exhausted humans had to barter with food to get him off.

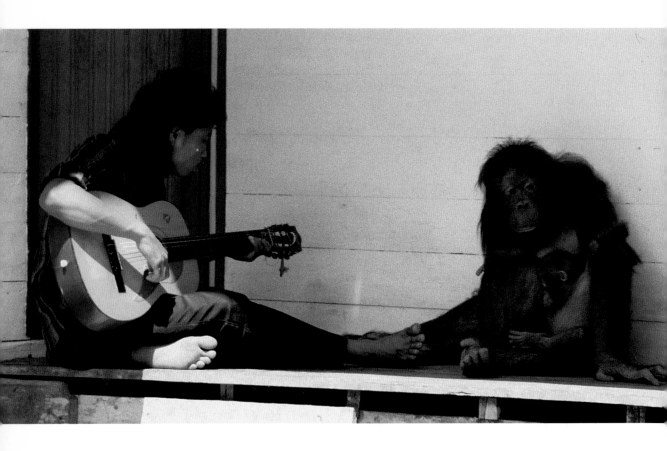

Ex-captives enjoyed being entertained. Supinah reclines and listens to Suto play the guitar.

another. They mated and reproduced as effectively as wild orangutans, indeed even more often, and they mated with wild orangutans as well as with their fellow ex-captives.

Orangutan males are notorious for having a fuzzy sense of species boundaries, and have actually kidnapped human females for sexual purposes, but some orangutan females share the tendency. Siswi, for instance, had a bit of an eye for humans. Connie Russell, doing her thesis research in the park, met Siswi on her first trip to Camp Leakey with me. Siswi played hostess perfectly, greeting Connie, taking her hand, and spending hours with her on the bunkhouse porch. Connie was dead charmed and put Siswi to the top of the must-meet list for her husband, John Ankenman, who was to arrive a few weeks later. When Connie brought John to camp, she was tickled to see Siswi and even more pleased when Siswi took John's hand and led him down one of the paths. She was pleased, that is, until she tried to join them. Then she got the boot. From Siswi. Siswi wanted John, and only John, to stroll with her through the forest. Siswi flirted, she lolled about, she tried to pull John to the ground to wrestle with her. All Connie could do was trail along the proverbial seven steps behind. Any time she tried to join the twosome, Siswi maneuvered her away. Nor were John's efforts to extricate himself any more successful. Whenever he tried to extricate himself, Siswi just tightened her grip. Connie and John finally managed to escape by distracting Siswi with an offer no orang-utan can resist: food.

Other than mastering forest life, the big item on the rehabilitation agenda was getting ex-captives to abandon human ways. Some ex-captives did so, but others apparently saw little merit in the idea. If human practices made life easier, why not keep both? On top of the attention, protection, and support that humans provided, they added variety and spice to life. Human goods added a certain luxury to life. Ponchos gave better protection from rain than leaves could ever do, hammocks swayed so restfully, sign language made for more effective begging from humans, and human junk made great toys. Nancy, for example, swiped empty rice sacks, took them up trees, hooked them over broken branches, dived in head first so her feet stuck out like a frog, then twirled and squirmed inside. The human world just offered more scope.

For all that, once I got to know the rehabilitants, my sense was that we humans weren't very important to them. My impression was that they thought we were cats—figuratively speaking, of course. They were well aware that we were there and they sometimes found us amusing or somewhat of a nuisance, but for the most part we were just a rather unimportant part of the background. They worked the peculiarities of the human world as they'd probably work the peculiarities of a swamp or a hill forest, as just one more slice of their complex world.

HE SOR-CERER'S RED APPRENTICE

My official reason for being at Camp Leakey was to conduct research on imitation, an important area of study at that time for anyone interested in non-human intelligence. Since the turn of the century science had accepted what non-scientific observers had long assumed, that many animals imitate. Parrots, for instance, parrot; and apes ape. In the late 1980s it was pointed out that there are many ways to create imitations, meaning copies of a model's actions, and that some of those ways are mentally simple, while others are more complex. Imitating an action one already knows, like yawning, is very simple. Even dogs and cats can do it. But imitating actions that are new to the imitator, that is, learning new actions just by watching someone else perform them, is not simple at all. This imitative learning is understood to require reasoning, and as such it has been considered an ability that humans alone possess.

So the many reports of great ape imitation that had accumulated over the twentieth century were discounted. Such imitation was assumed to be simple-minded rather than complex. I considered that this conclusion might have underrated great apes' potential because of a factor that had not been taken into account: most of those reports involved captives who could have been intellectually disadvantaged by

Orangutans, fringed with copper in the dawn mists, move through the trees.

Every morning for about fifteen years, Pak Akyar brought loads of food to the dock and doled it out to any ex-captive orangutans that wanted it. He made sure that each got his or her fair share, and adjudicated any disputes.

the privations of captive life. My aim was to explore imitative learning in the crossover human-plus-orangutan world of rehabilitation, an arena where the subjects of study had a double set of opportunities, a double set of models, and time to fool around. What I meant to do was follow the rehabilitants and look for instances of their imitating others' actions as well as other events that offered clues about how they could have learned those actions.

Finding orangutans to follow was easy because many Camp Leakey rehabilitants kicked off their day at the dock with the provisions served there at seven each morning. I often headed down early, because waiting there was one of Eden's delights, dawn's mists luring sunlight down milky shafts, gibbons' great calls soaring, and breezes sending downy wisps of coolness across your skin, as seductively elusive as fine perfume.

Back at the kitchen, Pak Akyar would be readying big pots of food for the orangutans, as he had for fifteen years. He'd be loading the pots on the camp's wooden wagon, along with any orangutans waiting to hitch a lift, then wheeling it all down to the dock. You knew he was on his way because he called each orangutan by name as he approached the base of the dock—"Princess, Unyuk, Siswoyo, Davida, Pola." Those in the forest who heard, and chose to come, emerged from the forest as great shadows fringed copper by the morning sun, or announced themselves by the swish and sway of branches bowing under their clambering. Those in camp ambled along the web of paths to the dock and any who had nested in the observation tower oozed over the tower's railing and down the side.

They shuffled up in ones and twos, according to family, friends, and rank. Sometimes as many as thirty arrived, sometimes only one or two. Siswoyo, the camp dowager, came with her infant son Sugarjito and often

Some of the orangutans seemed to shop for the week, gathering as much food as they could then retiring to a quiet venue to eat in peace.

Siswoyo was one of several rehabilitants intrigued by siphoning. She didn't quite "get it" though: here she's trying to insert her siphon into the fuel drum, but hasn't realized that she has to remove the drum's cap first.

her older offspring plus their sundry cronies. She sailed to the center of the food, chose what she wanted, and savored the eating. Lower-ranking orangutans, like Princess or Supinah, hung back waiting for Pak Akyar to deliver them a portion privately, or snatched a bit then beat a hasty retreat before Siswoyo arrived. Two wild orangutans who had discovered Camp Leakey years ago, Mooch, an adult female, and Ralph, an adult male, often joined the party. Mooch invariably arrived with a show of outraged kiss-squeaks, belches, and branch throwing. For everyone else, it was an occasion to play, meet, and relax on full stomachs. Ralph must have seen it as the land of Oz—free food every day and dozens of plump, healthy females. A couple of hours often passed before the scene broke up and orangutans headed for their daily rounds.

Then I had to gear up for following my orangutan of the day. It could have meant scrambling up trees and through swamps, but I lucked out: it was around camp, where they might scare up more food at the dining hall or entertain themselves with activities like sawing wood or putting on clothes. In other words, it was prime territory for imitation. If the orangutans imitated human rather than orangutan actions, that was an advantage, because I stood a good chance of recognizing it.

What I found was that imitation was almost everywhere. Many of the rehabilitants' oddball activities—riding boats, washing laundry, shooting blowguns—turned out to involve imitations of the techniques they regularly saw camp staff use.

Siswoyo gave me a good taste of copying. Well into her twenties, she was a resident of long standing, one of the rehabilitant veterans. I was told she'd arrived when she was five or six years old, rescued from a cage so cramped she couldn't stand up. Even now, after years of freedom, she couldn't straighten her limbs. Otherwise she'd thrived. She established her range around camp and bore three fine youngsters. Her copying included siphoning fuel, sawing wood (including inserting the saw into the cut the human worker had started), and "writing" in a notebook—always the same way camp staff did.

During one of the periodic bouts of cleanup around the camp, Mr. Mursiman, a long-time staff member, was cleaning paths by removing weeds that had grown in along their edges. He used the standard camp technique, slicing weeds at ground level with a hoe then tossing the cuttings into the bush. A very precise man, Mr. Mursiman piled his cuttings in a neat row behind him along the center of the path before discarding them. Siswoyo knew Mr. Mursiman well. When I arrived, she was close behind him on the path, also removing weeds from the side and piling the cuttings into a row behind her in the middle. Her row of cuttings wasn't so neat (in fact, it was pretty messy), she used a half meter long stick instead of a hoe, and she chopped weeds out roughly or pulled them out by hand rather than slicing them off cleanly. But it wasn't her fault. Mr. Mursiman had the camp's only hoe. When I asked Mr. Mursiman how it had started, he said that Siswoyo had followed him, watched his weeding, then started weeding herself.

Galdikas had mentioned Princess as someone worth knowing. Princess had been sent to Camp Leakey as a tiny infant and early on had become a pupil in Gary Shapiro's sign-language project. When I met her, she was a young adult with her first son, three-and-a-half-year-old Prince. She still used a few of her signs in her life around camp. She imitated all sorts of things: scribbling in books, weeding and sweeping paths, washing laundry, and picking locks.

She was one of the camp's finest lock specialists. Rehabilitants didn't voluntarily respect the no-breaking-into-buildings rule, so doors were all fitted with bar locks that shunted back and forth, through brackets, across the door jamb. Bars were mounted inside and a narrow slot was cut in the door as a "keyhole." To unlock a door from outside, one made a stick key, poked it through the keyhole, and jimmied the bar free of the jamb. The system was not orangutan-proof: Several had mastered the technique. Princess most often hit the provisions storeroom. She'd find a stick roughly the right size, bite it down to fit, poke it through the keyhole, wiggle it back and forth until the bar was freed, open the door, and help herself to the food. Once, on finding the door locked, she looked under a loose plank on the storeroom porch before searching for a key stick. At the time I wondered why. Later I discovered that Pak Akyar often stashed a key stick under that plank, probably so he wouldn't have to make a new one every time he wanted in.

Orangutan mothers like Princess and Siswoyo showed their expertise to their offspring and friends, who imitated it, then showed it to their friends, and so on. Youngsters imitated so commonly they were like their mothers' little shadows. One of Princess's storeroom heists netted long pieces of sugarcane. Usually orangutans got short pieces, cut up, which they ate by biting off the tough sheath then chewing the inner fiber for its sweet juice. But Princess held her long cane horizontally, grabbing one end in each hand, then pulled her two hands together sharply. The cane arched upwards into a bow then cracked in the middle, squeezing juice out at the crack. She simply opened her mouth underneath and caught the dripping juice. As Princess was in the middle of this job, I caught sight of her son Prince sitting on the ground nearby with his own piece of sugarcane, a little short one. He too held the cane by each end and bent it into a bow, but he held it in his feet and bent it by pushing against its middle with his hands.

Companions also imitated one another. Davida, a juvenile female, copied her favorite humans; Earthwatchers were among them. She came to the bunkhouse one morning and won a special treat, a toothbrush with toothpaste on it. Orangutans stole toothbrushes or toothpaste but rarely got both, especially combined this way. Brush in fist, Davida inserted it in one side of her mouth, closed her lips around it, and scrubbed the bristles back and forth over her teeth/tongue. That side done, she twisted the brush to the other side of her mouth and scrubbed there too. When she finished brushing, she climbed onto the porch rail and spat the used toothpaste over the rail onto the ground. Her technique was a dead ringer for the one Earthwatch volunteers used, especially spitting the used toothpaste over the bunkhouse rail onto the ground. Orangutans usually eat toothpaste; they don't spit it out.

I even found myself in the game, with Pegi, a juvenile female who had recently been sent for rehabilitation. Pegi wasn't coping. She spent her days huddled under buildings, hugged herself, wouldn't eat, and had developed diarrhea. I spent days helping her adjust to her new world and racking my brains for ways to tempt her to eat.

I finally came up with a flash of brilliance. Raisins! I was so sure she'd like them that I raided my private stash. But when I put one to her lips, her mouth clamped shut and neither persuasion nor force would open it. Seeing no point in wasting a perfectly good raisin, I ate it myself. She eyed me very closely as I munched, so I offered her another one. She ate it! Then she ate another, then several, then a handful. Within two weeks I had to stock my pack with two boxes of raisins daily. When she saw me coming, she tore open my pack and crammed all the raisins in her mouth, box and all. She must have refused the first one because she didn't know what it was. Now, she clearly loved them. Before she'd try a new food, though, she had to see me eat it: she *had* to watch to learn.

Pegi then turned the tables. I discovered she liked the tender white tissue at the base of blades of grass, and I spent a half hour one day pulling blade after blade of grass for her as she lounged, princess-like, in my lap. Finally tired of the job, I turned away to watch staff play volleyball—until I felt a slight tickle at my chin, looked down, and saw Pegi looking solemnly up at me, holding the base of a blade of grass to my mouth.

It was Supinah's imitations that topped the heap for complexity. Supinah was another young adult female and she was one odd bird, one of those orangutans that wouldn't be. Despite rehabilitation she worked so hard at mastering human ways that camp residents once almost voted her an honorary human. She put more effort into washing laundry and sawing logs than she did into orangutan skills. When she ventured into the forest, she broke the trees.

She was a prodigious imitator: she hammered nails, sawed wood, sharpened axe blades, chopped wood, dug with shovels, siphoned fuel, swept porches, painted buildings, pumped water, blew blowguns, fixed blowgun darts, lit cigarettes, (almost) lit a fire, washed dishes and laundry, baled water from a dugout by rocking it side to side, put on boots, tried on glasses, combed her hair, wiped her face with Kleenex, carried parasols against the sun, and applied insect repellent to herself. Whenever the job involved a complex technique, hers matched the one used in camp.

Supinah's crowning achievement had to be making a fire. Almost. Making fire is regarded as a significant leap forward for the human species that we believe uniquely ours. So Supinah's near-success is outstanding. It happened like this. Supinah wandered into the camp's outdoor cooking area near midmorning, at a rare moment when it was left untended. Breakfast cooking fires were still smoldering and all the cook's gadgets were nearby—firewood, a container of kerosene, a plastic cup for scooping kerosene from the container, and an aluminum lid. To start fires, the cooks usually soaked a stick with kerosene then lit it with an

already burning stick. To make fires burn hotter, they blew on them or fanned them with the aluminum lid, held vertically and waved briskly from side to side.

Supinah first picked up a burning stick. Then she went to the container, removed the plastic cup and aluminum lid sitting on top, scooped out a cup of kerosene, and plunged the burning end of her stick into it. I thought the cup and container held water, not kerosene, and announced that Supinah was trying to put out her fire. Jane Fitchen, the

One morning when I was feeding Pegi blades of grass, she turned the tables and started giving them to me.

77

Whenever Supinah could pilfer tools hammers, saws, axes she would enjoy helping with camp construction work. She once managed to saw three-quarters of the way through a log ten centimeters thick.

student helping me, murmured that Supinah was trying to *start* a fire, but I was fussing with my camera, paid no attention, and let Supinah go her merry way. Fortunately for us, plunging the stick into the kerosene put the fire out.

Supinah was determined to light her stick again and tried all the operations and gadgets the cooks used to light fires: wetting the stick with kerosene, blowing on its tip, touching other burning sticks to it, and even fanning over it with the aluminum lid the same way the cooks did.

She kept trying for fifteen minutes before she tossed her last stick into the embers, poured out the kerosene, and left. Later, talking about Supinah's fire works, Galdikas recalled having once caught Supinah trying to torch the dining hall. At the time, she said, she'd assumed Supinah had found a stick already burning. After what Jane and I had seen, it seemed entirely possible that Supinah lit the stick herself.

I also saw Supinah siphon from a fuel drum into a jerrycan, the same way camp staff did. The fuel drum was empty when she tried though, the

Several orangutans understood brushing tools and tried them out when they got the chance. They used them correctly combs and hairbrushes for hair (although they sometimes combed in the wrong direction), scrub brushes for scrubbing floors or laundry, and toothbrushes only for teeth.

Above: Supinah came into the camp's outdoor cooking area, found a piece of burning wood, and went to get kerosene from the cooks' can.

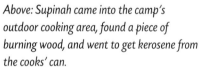

Right top: Supinah pours some of the kerosene into her little plastic cup.

Right centre: Supinah puts her (doused) stick into the cup of kerosene and picks up a round aluminum lid to fan over it.

Right bottom: The cup of kerosene and fanning didn't help, so Supinah takes the stick out of the kerosene and blows directly on its tip, where it had been burning before.

only condition under which staff tolerated orangutans' fiddling with fuel drums. The reason should be obvious: no one wanted to donate a month's supply of fuel to Supinah's experiments on fluid dynamics. The camp siphoning technique was the same as the one most of us know. You open two containers, a fuel drum (with fuel in it) and a jerrycan (to receive fuel from the drum), place the jerrycan near the drum, insert the first end of a hose into the drum all the way into the fuel, suck the second end of the hose until fuel flows through it, then quickly insert it into the jerrycan. If you get it right, the fuel flows.

Supinah opened both containers, placed the jerrycan near the drum, inserted the first end of the hose into the drum and the second end in her mouth, bellowed her cheeks out (I couldn't tell if she sucked or blew), then inserted the second end of the hose into the jerrycan. Since the drum was empty, she missed a few fine points—like the first end of the hose had to be in fuel, not just in the drum, the hose had to be sucked, not blown, and it had to be sucked until fuel flowed. She seemed to grasp that something had gone awry, though, because she pushed the first end of the hose farther into the drum then pulled it out it and tasted it, and she tried to drink from the (empty) jerrycan. Then she started again: she put one end of the hose into the drum and held it in place with a foot, put the second end of the hose into her mouth, bellowed her cheeks, and stood the jerrycan up. In the process she dropped the second end of the hose, so she had to pause to retrieve it before inserting it into the jerrycan. Of course she never siphoned any fuel but her form was close to impeccable. That fact makes the case for imitation even stronger: if she'd never been able to practice real siphoning, how could she have learned to do it—except by watching?

If Supinah's imitations were the most complex, one other took the cake for breadth. It was a special way of eating soap. Many species savor soap, so the fact that ex-captives ate it was nothing special. What was special was that they all did it the same, intricate way. They took a piece of soap, wet it, rubbed it into a lather, then ate the lather. To rub, they

Herbie, now an adolescent, is an expert soap latherer.

put wet soap on the back of one forearm, grasped lightly around the forearm and over the soap with the other hand, then rubbed the soap up and down over the forearm hair. Once they had a thick lather, they licked it off their forearm as if it were gourmet mousse. I've kept my eyes open over the years to see if this happens elsewhere. Other orangutans certainly eat soap but none have ever used the Camp Leakey method. Some just work up suds in their mouth. Some do make a lather on their arms but without the grip used at Camp Leakey. If I'm right, then this was a community tradition of the sort now being seen as the seeds of culture.

As far as I could see, at least some of these skills had to have been acquired by imitative learning, at least in part. No one would have taught the orangutans these skills. Galdikas didn't want them weeding paths, because of the mess they made. Nobody wanted orangutan-made pancakes, probably not even orangutans. And nobody in their right mind would teach a free-ranging orangutan to start a fire. In fact, everyone discouraged the orangutans' learning these skills; they even hid the tools. Slip-ups did occur, which allowed the orangutans occasional chances to practice on the sly. But orangutans are slow learners compared to humans, and a few sly chances are not likely to be enough for the acquiring of complex skills. Even if they were, observation of humans obviously played a part. After all, what's the likelihood that fiddling about would result in the very same, often unusual, technique that camp staff used?

These cases all convinced me that orangutans are capable of learning by imitation. They may not do it exactly the same way as humans do, or with the same power, but they do it. If imitative learning is the ticket that allows crossing Aristotle's Rubicon into the realm of reason, then for my money their ticket is valid.

Mechanical genius

The complement to gaining expertise by imitating is figuring it out on your own. If rehabilitants used imitation to advantage, their successes stemmed in large degree from their understanding of how things work. Supinah came close to making a fire because she imitated but also because she improvised sensible "fixes" when her copy went wrong. To do that, she must have had a pretty good understanding of how fires work. Think about it. She failed because of only one error, how much to wet sticks with kerosene. It's a subtle point that foils many human attempts to start barbecues, so she did no worse than many of us. I decided to take a close look at the ex-captives' understanding of tools.

Tools are hot items in the study of intelligence. Technically speaking, they are external, detached objects an individual uses to assist in reaching a goal. Not all tool use shows high intelligence. Birds use twigs as probes and otters use stones to crack oysters, but neither skill bespeaks great wit because their usage is unvarying—narrowly limited to one type of tool which is used for one purpose, and performed in a rote, mechanical fashion. But using tools widely and flexibly is seen as a mark of high intelligence, and making tools an intellectual cut above that. An intellectual capacity for tool use and manufacture is considered to be an evolutionary adaptation—just like the acuity of the hawk's eye and the speed of the gazelle, shaped by evolution to serve survival.

Others had of course studied tools in orangutans, so we already know that captive and ex-captive orangutans may be the most sophisticated tool makers of all the nonhuman primates. The ex-captives' exploits with tools were worth exploring, to sort out how complicated their understanding could be.

The number of different tools rehabilitants could combine in one procedure gives a good indication of the complexity of their understanding. Supinah's fire-making involved using five different tools together. Pancake making involved combining eggs, flour, a cup, and a

spoon. Princess also created an elaborate combination to cool some scalding hot coffee she had in a plastic bottle. She retrieved an old soya sauce bottle, bit off its cap, and poured the coffee from her plastic bottle into it. Princess's work was a bit messy, with some of the coffee spilling down the side of the soya bottle, but most of the coffee did pour into the soya bottle and the pouring did cool it enough to drink it. Her pouring coordinated four different items—two bottles, the soya bottle's lid, and hot coffee. And given the soya bottle's narrow neck, the intricate understanding she showed in pouring was none too shabby.

Some tool-using complexities were more subtle. Princess once used a burning mosquito coil as a substitute "pen" to "write" on a piece of paper. She didn't just stupidly fail to notice the difference between a pen and the mosquito coil—she didn't scribble with the coil the way she scribbled with pens and she "fixed" the mosquito coil by blowing on its burning tip, not the sort of fix she'd ever attempted with a pen. Instead, she recognized the two worked in similar ways, an impressive bit of logic.

One way of making intellectual sense of the orangutans' tool work is to consider the development of similar skills in human children. Princess's handling of hot coffee involved many different linkages between tools and other objects; it was a complex task that a human child would not be able to manage before about two and a half to three years of age. If that sounds unimpressive, remember that human three-year-olds have a considerable command of language, rudimentary plans, and rudimentary ideas. In other words, they're little thinkers. So orangutans like Supinah and Princess are thinkers too.

Kusasi, thinking things through

Princess has found a piece of burning mosquito coil, and has used the tip to write on the piece of paper at her feet. Here, she blows on the burning tip to enliven it. Her daughter Peta, about two, studies the situation intently.

Plans

I gained insight into other facets of orangutan minds by accident. Among the most impressive ability I detected was planning. Despite old views that animals are locked in the here-and-now, the suspicion has lurked for years that great apes plan ahead. The best plan I encountered involved an orangutan-human encounter. Humans may have recognized the plan because, in a sense, it was played out on our turf.

Princess was at her social best one afternoon, visiting old-timers and new volunteers at the bunkhouse porch. She was always a favorite with visitors and she duly charmed these newcomers too. After a while she had other things to do, so she left, orangutan style, up the building to the peak of the roof. Unusually, she stayed up there only a few minutes, and then instead of continuing on her way she returned to the porch. Within minutes of her return, even old-timers were exclaiming on how especially sweet she was. She liked humans so much that day that she stayed past four o'clock, when she normally went for provisions, past sunset at six p.m., when she should have made a nest, and on into the night. She even slept across the bunkhouse doorstep with her daughter Peta, who was just one year old. At eight that evening, when the dinner bell was rung, people had to step carefully over the sleeping Princess and Peta when they left the bunkhouse.

Humans came back after dinner, a couple of hours later, to find Princess gone. She'd probably left for a nest, as is proper and good, they thought, as they unlocked the bunkhouse door. But they couldn't get in, the door wouldn't budge. Someone shone a flashlight inside. And there was the *inside* bar lock, closed, and Princess and Peta sitting quietly on the floor. Princess was the only one who could lock or unlock the door so one old-timer asked her, in Indonesian, to open it. She obligingly did so then walked calmly out, past astonished humans. Inside, people searched for damage and for an explanation of Princess's reverse-Houdini act.

Damage wasn't bad. The worst was a bottle of Pepto-Bismol poured into a can of Gatorade. It was upstairs that someone finally spotted how Princess must have broken in. There was a hole at the top of the end wall, at the peak of the roof.

Thinking back, people realized that the hole was where Princess had paused a few minutes that afternoon, and that it was just after going there that she began to act strangely. It adds up: she had spotted a place to break in, but, realizing there was no hope of a successful heist with humans around, she stayed nearby until she knew they were gone for a long time, then made her hit. In other words, she planned the break-in hours ahead. Patience is a well-known orangutan virtue, so this is well within the realm of possibility. What is even more humbling for the humans involved is Princess tricked them. She acted uncharacteristically nice while simultaneously showing many signs of waiting for something. None of her human dupes spotted the ruse, not even those who knew her well.

Abilities in concert

Sometimes several abilities played together into one symphonic event. One of the best involved Unyuk, an adult female. I followed her rarely, although she was energetic, funny, and inventive, because she was also wickedly feisty. She was notorious for thwacking people and orangutans with big sticks. One student had tried to handle her by turning the tables and waving a big stick at *her*. Her response was to get her own stick and wave it at him. The student fled right out of the forest, Unyuk in hot pursuit.

One day when Ucing, a field assistant, came along, I braved following Unyuk. For something to do while I filmed, Ucing used his Swiss Army knife to made a walking stick. He sawed off the stick's ends and scraped off its bark. Unyuk watched closely then imitated Ucing's sawing and scraping with two sticks, one against the other. Probably because Unyuk was so interested in the knife, Ucing opened its scissors and pretended to snip her hair near her forehead. Unyuk cringed nervously but when Ucing paused she looked at him, grabbed a fistful of hair from her head with one hand and both her sticks with the other, then made a cutting motion with the sticks across the fistful of hair. Unyuk's gestures prompted Ucing to snip at her hair again, at the spot she indicated.

I was in the midst of videotaping this when I suddenly got the feeling she wasn't interested any more. Later, studying the video, I found that Unyuk's eyes had suddenly stopped tracking the knife's movements and were staring beyond Ucing to where our backpacks lay on the ground, unguarded. But Unyuk acted as if she was still involved in scissors play, continuing to offer her arm and head to Ucing to snip. Twice she moved, as if shifting to a more comfortable position, but each shift was to the right and after the second, she was beside Ucing rather than in front of him. Within seconds she lunged for the packs. We just barely saved them.

So here we had imitation, complex gestural communication, tool use, and deception, each a highly sophisticated ability, all combined together to make one complex event. I don't have a way to put a number on that complexity, but it was sophisticated enough to fool a human who'd known Unyuk for six years.

TROUBLE IN EDEN

Camp Leakey's unusual world disintegrated over the years I visited. The immediate cause was disputes that flared over rehabilitation, but, in fact, the disputes had probably been brewing for years. Unfortunately they weren't just political squabbles peripheral to orangutan affairs. The ex-captives were what the disputes were about, and the outcome of the disputes would determine who controlled their lives and how they would be allowed to live. Of all the things that went wrong, two deserve mention: ecotourism, the heart of the problem, and the incident of the "Taiwan Ten," which precipitated the disintegration.

By the end of the 1980s, up to a hundred Earthwatch volunteers visited Camp Leakey yearly; returning ex-volunteers and individual visitors swelled the numbers. Indonesian guidebooks began promoting Camp Leakey's orangutans and giving directions to the park, providing an open invitation to any reader inclined to show up. As the camp's fame grew, film crews, government officials, and film stars joined the influx. Kelotoks jammed the river, and the camp overflowed. Sometimes fifty people crammed into the dining hall for dinner.

Sometimes so many visitors arrived at Camp Leakey at the same time that the dock overflowed with people and there were kelotok jams on the river.

Ecotourism was supposed to contribute to conservation by giving the park and its orangutans economic value as visitor attractions and by educating visitors to greater appreciation of nature and involvement in

Yuni and her year-old infant are hairless, emaciated, and itching from a skin disease that spread among the rehabilitants in 1993. Yuni had so little hair that her infant could no longer cling to her.

protecting it. Earthwatch teams assisted conservation-oriented research and were educated in the process. For instance, Connie Russell, from the Faculty of Environmental Studies at York University in Toronto, came to Camp Leakey in 1992 to assess the educational rationale for ecotourism. She studied what Earthwatch volunteers learned about orangutans.

Most volunteers had devoted substantial effort to gathering information about their trip, so they arrived with quite well-developed notions about orangutans. Over their two weeks at Camp Leakey, however, with all its educational programs, they pretty much stuck to their preconceptions.

Some volunteers imagined young ex-captive orangutans to be much like human infants and sought every opportunity to cuddle them. While these young orangutans undoubtedly needed comfort, the volunteers seemed unable or unwilling to grasp that they could infect the orangutans with human diseases or that the orangutans could injure them. Other volunteers fixed on wild orangutans to the point of dismissing ex-captives as not "real" orangutans. They spent their days in the forest seeking wild orangutans, putting pressure on the wild areas. Another set of volunteers appeared to view their entire trip through a lens. They rode up the river, camera never leaving eye, waiting for the perfect shot—but missing all the beauty, subtlety, and complexity on the periphery.

Connie found that ecotourism is not automatically educational, may not contribute to conservation, and can create problems. Ecotourists gave rehabilitants opportunities to develop their talents as pickpockets and moochers. New visitors were perfect dupes: naive, timid, and insufficiently vigilant. They were also easy marks for rides or cuddles; some youngsters would simply climb up their chosen supporters' legs into their arms or onto their backs. Food sharing and cuddling opened the

way for human infections to spread to the rehabilitants, and from them perhaps to wild orangutans. Rehabilitants were indeed falling ill, and Galdikas remarked that it seemed to be the most cuddled youngsters who were most often stricken. Several of them died.

Romeo was the cuddliest of the Taiwan Ten.

The affair of the Taiwan Ten began late in 1990 and came to a head late in 1991. It brought Galdikas's problems to crisis level. The Taiwan Ten were the second group of ex-captive orangutans confiscated internationally and repatriated to Indonesia, a mere ten of hundreds smuggled into Taiwan in the mid 1980s to fuel the exotic pet trade. The first repatriated group, the Bangkok Six, consisted of six infants seized from smugglers at the Bangkok airport in 1990. They'd been sent to Galdikas for rehabilitation but fell ill within months. All but one had died or disappeared within two years. Galdikas wanted to take charge of rehabilitating the Taiwan Ten too, but the Indonesian government planned to send them to the Smits rehabilitation facility at Wanariset.

Galdikas is not one to give in. She took control of the Ten. She moved them to a house she had rented in Jakarta and had her supporters, mostly university students, care for them. I visited the Ten several times at that house and got to know some of them—Grinch, a little male and a favorite with cuddling visitors; Manis and Imelda, females, a shy little slip of a thing and a feisty thug; and Charlie and Romeo, the largest males, one too tough to handle and the other fat and bald but the greatest cuddler of all.

This situation became a year-long standoff. Late in the summer of 1991, officials made a surprise visit to the house to test the Ten for disease. Their results identified three carrying infectious disease—two with tuberculosis, one with hepatitis B. It was Romeo, the great cuddler, who registered hepatitis B. All ten had been certified healthy when they arrived from Taiwan so each side held the other responsible for the infection. Galdikas got her side of the story to the international media. That pressured the government but hardened its attitude. Finally her students "kidnapped" nine of the Ten, whisking them through the labyrinth of

Jakarta's streets to escape confiscation. Nine? They left Charlie behind, the one orangutan too rough and tough to handle.

That was, essentially, the end of the affair. The students negotiated a truce and handed the orangutans over, the seven healthy orangutans went to Wanariset in early November, and the three infected ones went to the Indonesian Primate Institute. But its repercussions reached Tanjung Puting. By the end of the year, park authorities had taken control of orangutan rehabilitation and all visitors in Tanjung Puting, including visitors to Camp Leakey. By 1992, no visitors, not even Earthwatch volunteers, were allowed to stay in Camp Leakey or anywhere inside the park without a formal permit. That was the last year that Earthwatch sent teams to Galdikas's orangutan project.

But other, broader factors were also working against the rehabilitation project in Tanjung Puting. Many specialists in orangutan conservation had been monitoring the results of the orangutan rehabilitation programs established in the late 1960s and early 1970s, including Galdikas's, and ever since the late 1970s they had been advising changes to rehabilitation practices.

The original principles of rehabilitation were, first, that ex-captives should be reintroduced to wild orangutan populations to restock them; second, that visitors to rehabilitation projects should be encouraged for the educational value; and, third, that re-education of ex-captives should focus on forest rather than social skills (because orangutans are essentially solitary). Experiences at many rehabilitation sites had demonstrated that what sounded good in principle did not work so well in practice. Ex-captives probably overcrowded wild populations, rather than revitalizing them, given the rate at which orangutan habitat was disappearing. Social skills turned out to be difficult for ex-captives to learn, perhaps more so than forest skills. As for visitors, it was well known that they undermined rehabilitation by encouraging ex-captives to become moochers rather than forest dwellers, and that they introduced real dangers in the form of human disease.

Considered opinion was that the original principles, and the programs that still adhered to them, should be scrapped. New principles were recommended. Ex-captives should be carefully screened and treated for disease; human contact and human dependency should be firmly curtailed; orangutan social relations as well as forest skills should be fostered; and ex-captives should be released only into areas free of wild orangutans. Galdikas's rehabilitation program sat staunchly in the traditional mold.

I observed all this from Galdikas's camp at the time, and so I saw the criticisms and restrictions as ill-founded predatory attacks. Later, as I developed a broader view of orangutan protection issues, it was increasingly clear that the criticisms were justified and that the problems generated by Galdikas's style of rehabilitation outweighed the benefits. It was indeed time for the world of Camp Leakey to change.

It was also time for me to change. With the political uproar, it became impossible for me to continue my research at Camp Leakey. My interests were also changing, and turning to how orangutans use their minds in the forest. It would take forest-minded orangutans to show me that, not the bicultural ones of Camp Leakey. I began looking for another research site. Timely or not, it was with deep regret and a sense of enormous loss. The understandings and relationships I had worked out with these orangutans over four years had been invaluable to me, however inconsequential they were to them. I could never have followed them so closely if I hadn't established a positive relationship with each one.

The value was personal as well. It had been such a pleasure to step onto the Camp Leakey dock in 1994, after a year's absence, and have Princess immediately recognize me in a lineup of Western tourists. She came up, grabbed my hand, and took me for a walk with her. Not to exaggerate, I suspect she recognized my red pack, not my person. I was still charmed. Moving meant building all that anew.

ANARISET: THE NEW REHABILITATION

W here I ended up in 1994, after casting about for other sites, was at Willie Smits' Orangutan Reintroduction Project (ORP) at Wanariset. He freed ex-captives in a forest where no tourists or camp activities could distract them from readapting to life in the forest.

Wanariset was no tranquil forest post, easygoing and slow-moving like its loitering apes. It was a modern research institute run jointly by the Indonesian Forestry Department and the Dutch Tropenbos Foundation, and it specialized in tropical forests, not just orangutans. It was on a main highway just 40 kilometers from the oil city of Balikpapan, not deep in the rainforest. Willie was not an orangutan or a primate specialist, but a microbiologist specializing in tree-fungus interdependencies, and the team leader for the entire Wanariset project. He had mounted the orangutan project as a personal venture in 1991, with the help of children from Balikpapan's international school, and installed its intake facilities at Wanariset.

Willie had set up his program for ex-captives on the model recommended by Herman Rijksen, one of the experts who had worked on redesigning rehabilitation to avoid the pitfalls of traditional practices. His realization of Rijksen's design included strict disease man-

As you walked past cages at Wanariset, dozens of hairy red arms reached out.

agement under formal veterinary care, mammoth "socialization" cages in which healthy ex-captives were housed together to turn them towards orangutan social life, and a minimum of human contact. Tourism was sharply restricted, and ex-captives were released in groups, into forests with no wild orangutans. In 1995, this became the only legal method of dealing with ex-captive orangutans in Indonesia.

The love driving this program was tough love, not mother love. If Camp Leakey was a world without barriers, ORP set them firmly back in place. From the moment orangutans arrived at ORP until the moment they were released into the forest, they were separated from humans in cages. When I arrived, those cages overflowed with damaged orangutans of all shapes and sizes. They plastered themselves against the bars when anyone passed by, arms groping frantically out. One cage kept a quintet of fragile infants under Plexiglas, to bar human contact. A second, just across the walkway, held a dozen older infants. They had an air of health with their goofy, exuberant tumbling about but most were almost hairless. A third, downhill, held dozens of juveniles. Around a corner was a row of stall-like cages for isolation, each holding one waif. Each looked out on a wall and barriers prevented contact between them. It looked like, and basically was, an orangutan refugee camp.

As you'd expect, as a primatologist aware of the psychological damage that can result from life in sterile cages, I was distressed at this scene. A big part of the problem was the numbers: no one had expected this many orangutans. Willie had started out confident that one blitz to confiscate every captive orangutan in East Kalimantan would once and for all wipe out the illegal trade. Rijksen, who was visiting Wanariset when I arrived, remarked as we rode in from the airport that ORP should be able to close its doors within two years if it was doing its job properly. The reality was that Willie had already taken in nearly 120 ex-captives and ORP had been open less than three years. He had released about forty ex-captives into a nearby forest but there were still about eighty waiting in the cages and more were arriving every day. He was expanding the caging but couldn't keep up with the influx. Wanariset was on the point of overflowing.

The Taiwan Ten strike again

Despite the cage culture, it was clear by day two that I was back in the land of free-wheeling rehabilitants. The phone rang: it was a nearby plywood factory requesting help. There was an orangutan mooching at their canteen; it had been fun for a while but wasn't any more. Could someone come and remove the orangutan? Slick mooching spelled ex-captive, so it had to be one of ORP's. The project's vet, Kris Warren, invited me along. With transport cage, darting equipment, and Udin, one of the clinic technicians, we left to stalk the moocher.

The trip to the factory took us up river then to the edge of a forest near the canteen where, people said, the orangutan often appeared. Within minutes we spotted someone making a nest atop a nearby tree in full view of everyone, now and again peering down at us as we peered up, obviously unconcerned about the audience. As we debated how to stage a capture, Udin took a close look and recognized Otong, a female released over two years earlier. I looked myself and recognized Imelda, thug of the Taiwan Ten. We were both right. Names had been scrambled when seven of the Ten had been sent to Wanariset, so my Imelda was their Otong. The good news was that she had survived successfully: she was sleek, plump, and carefree. The bad news was that she'd found her way back to humans and invited herself back in.

By the time we revved up our capture operation, Otong was at the edge of the woods atop a grassy slope behind the canteen playing to a crowd of several hundred factory workers. She peeked coyly from behind trees then pirouetted excitedly, or ventured a few steps

onto the grass then skittered back to safety in the trees. Kris said darting was impossible: Otong was too far away and she couldn't be sure of hitting her safely. We had to get her closer, into the open, and on the ground.

We hung a bag of bananas on a low branch of a tree, just out of the woods and in darting range. Otong couldn't resist. Within minutes she sneaked to the lure. Just as she reached around the tree into the banana bag, Kris fired a dart into her forearm. Otong immediately fled. She could have disappeared—the drug can take minutes to act—but our luck held. She dropped to the ground asleep in 30 seconds. The crowd poured in and paraded behind us as we carried Otong to a transport cage for the trip back to Wanariset. After two years of successfully living free in the forest and what must have seemed to her like a perfectly reasonable venture into human land—she had been welcomed and encouraged, in fact—Otong was back in captivity. For her first few days at Wanariset, she was one very dejected orangutan. She couldn't know it, but she would return to the woods once they'd found a place to release her farther from humans.

Otong turned out to be the only forest orangutan I met that year. The orangutans' release forest, Sungai Wain, was not yet set up for research and Willie needed help enriching life in the cages. His program focused primarily on the medical and social side of rehabilitation but he needed to foster forest skills too. He did have leaves and local fruits brought in several times weekly. He had also had supposedly "orangutan-proof" playthings and climbing structures installed in the socialization cages. I saw none because the socialization groups, twenty to twenty-five orangutans all living and deconstructing together in one cage, demolished anything within weeks of installation. The only things left to play with were cement chips from the sidewalk, fallen twigs and leaves, and food remains. So my first months at Wanariset were as an orangutan occupational therapist.

Otong peeks out from the woods as we prepare to catch her.

Close-up on captivity

Stationed at the front lines instead of in the woods, I got a close look at what captivity offers orangutans. A few times I accompanied confiscation teams as they followed up tips on illegal captives. On one day alone a team confiscated four orangutans, all under four years old. Three were in one riverside village, kept as pets or for sale. Two were in makeshift cages too small to stand up in. The third, an infant under a year old, was stashed alone, in the dark, at the bottom of an empty oil drum. Even if the "owners" had honestly tried to provide good care, and some probably did, they knew little of orangutan needs and lacked the resources to meet them. They were poor villagers, barely able to feed themselves, so they probably fed orangutans only scraps.

Mostly I saw captivity's after-effects. My enrichment job took me into quarantine and clinic areas as well as socialization cages, so some of those after-effects were as recent as yesterday. Quarantine was stop one for all new ex-captives, for the tests meant to safeguard everyone's health. Clinic was stop two, for follow-up treatment.

I might open the clinic door on an operation to set a broken limb, an orangutan struggling against testing, or an autopsy. Damage ran from physical scars to infection. One orangutan was named Rantai, Indonesian for "chain," because she arrived at Wanariset still attached to one and sported chafing scars. Maiming and crippling were almost ordinary. Mojo's right hand, or what remained of it, was a three-fingered claw locked into a permanent crook. Others had suffered internally from abysmal care; it left Sebulu irreversibly blind and Boneng emaciated from starvation. Infections ran the gamut from diarrhea and scabies to virulent diseases like tuberculosis and several forms of hepatitis. Parasites were simply ordinary. Some of the problems were easily cured but in other cases orangutans were at death's door when they arrived. Some died within days despite medical efforts, and others faced years of treatment with an uncertain prognosis.

Behind the clinic area were quarantine facilities. First came a row of miniature wooden cages for the smallest infants. Runny noses, hacking coughs, parasites, diarrhea—you name it, they had it. They lay shivering or listless with whatever miserable affliction was blighting their tiny, tragic lives, and they withdrew, huddled, or cried desperately from the isolation. Beyond that was a ward of concrete-and-iron cells for older isolates. It held Panjul, a male pushing adolescence, with eyes of steel and a chip on his shoulder. He had been transferred from an older rehabilitation program, Bontang, where he had reportedly been a great tourist favorite. His problem was probably the stew of infections they'd passed to him. Way in the back was Somad, a tiny infant, stuck far from everyone else because he had a fiercely contagious illness. Miraculously he kept pulling back from the edge of death.

Also back there were two famous players. One was Romeo, the Taiwan Ten's most affable ape. Wanariset had freed him from the Primate Institute but it had not managed to eradicate the risk that he carried hepatitis B. He looked perfectly healthy but his tests consistently registered chronic infection so he could be contagious. That consigned him to isolation. The other was Daidai, a victim, like Romeo, of the mid-'80s fad for orangutan pets in Taiwan. At six months of age, Daidai had become a cherished member of the Cheng family in Taipei. With them she lived a life of piano playing, cars about town, and restaurant dining. She understood spoken Chinese and could converse a little herself with a set of idiosyncratic gestures for things like "please" and "thank you". She was no Mozart and no scholar, not even an orangutan genius, but simply an orangutan raised as a human in a human world. Inevitably, she grew difficult to control. For a while the Chengs managed with an electric prod but finally, when she was about seven, they gave up and agreed to send her to Wanariset for rehabilitation. The move didn't free her of human interference. First she was drafted to star in a documentary on the plight of orangutans, then Wanariset stopped her in her tracks. When Wanariset screened her at intake, she tested positive for tuberculosis. Daidai looked perfectly healthy, like Romeo, and chances were high that the positive result was false, but who would take the chance? She was detained indefinitely in quarantine, months at least, until treatment and more tests could prove her free of that disease.

My expertise lay on the psychological side, and there I could see equally serious scars. It's well known that if infant primates are isolated for long periods, deprived of comfort and contact, they develop autistic-like behaviors like self-clinging, stereotyped rocking, and excessive fear of new situations; if the isolation continues too long, the disturbances become permanent. Ani, who was about two when I met her, looked like a classic case. She clung to herself and rocked frantically at the least disturbance, and had apparently done so since she arrived as a tiny infant. Holding her just made it worse. She had been placed with other orangutans her own age and sometimes she seemed better but even going to the playground distressed her so much that she never left the sidewalk.

Herein, from my psychological perspective, lay another dilemma

Pur, only a few months old, was quarantined after a confiscation team rescued him from a village. He was in a bucket being readied for the soup pot; his mother had already been eaten. Regulations required him to remain in isolation until he tested free of contagious diseases. Even after he regained his health and was allowed contact with other orangutans, Pur would be dependent on humans for four or five years before he was mature enough to be released in the forest.

Siti, an infant female, was in a large social group of similar-aged orangutans learning how to associate with her own species instead of with humans.

for the ORP program. Its focus on medical management, especially quarantine with its potentially lengthy isolation and lack of stimulation, could generate its own evils. It could create casualties like Ani. It could also delay re-education past the age at which it could readily be assimilated. At least some situations could be adjusted to balance medical with psychological considerations, such as quarantining infants in pairs rather than alone, and providing them with soft items to which they could cling.

Socialization

Many psychological problems remained hidden until orangutans were healthy and living in the relative freedom of the socialization cages. Even there, these problems were hard to spot because they were embedded in the buzz of activity. The two largest socialization cages held hordes of rambunctious kids overflowing with energy. There were twenty-five juveniles between four to seven years old in one cage and sixteen older infants between two to four in the other. The cages were age-graded for the same reasons that schools are—so big ones can't bully little ones, so the right care can be provided, and so on.

A good deal of their social energy was productive. Each group had a role and rank structure that became evident when there was something to bicker about, like food or friends. It had bullies and weaklings, protectors and protected, social butterflies and isolates, leaders and followers, friends and foes. Paulie, a sweet and lively female, was everybody's favorite. Siti, a three-year-old female, was so interested in everything that she even braved her male cagemates, who commandeered control of any new activities, for a share of the action. Social roles were complex. A large male named Dingdong dominated the juvenile group. He not only won his own disputes, he also adjudicated others', taking sides if the dispute involved a friend and defending females from sexually harassing

males. Many normal orangutan social behaviors appeared, like begging for food, refusing to share, and withdrawing from conflict, so the ex-captives were indeed mastering orangutan social life.

In addition to the social pluses, these orangutans had their own inventiveness to see them through. One of the juveniles bashed a hole in the middle of his cage's concrete floor with cement chips scavenged from outside the cage. Others joined in. Normally only one orangutan bashed at a time, but there was usually a group monitoring the operation intently. When the worker dropped the job, a watcher took it up. Over about a month, they collectively enlarged the hole to a width and depth of half a meter. Another time Koko, a juvenile female, snitched a length of wire from a worker. She threaded one end of the wire through one link of a swinging chain then bent it double, like a long hairpin. Next she threaded both ends through a second link of the chain about fifteen centimeters from the first one, twisting the ends in and around the link so the wire was firmly fixed in place. That made a handle. She grabbed it, dangled, then spun as sweetly as a music box ballerina.

Psychological problems nonetheless lurked within the buzz. Many of those that surfaced looked like remnants of captivity. Their causes often remained obscure because "owners" disappeared, wouldn't own up for fear of prosecution, or casually dropped only snippets of information. Sometimes it was possible to make a good guess. Tono must have been trained to clown because he came with a bag of tricks like walking backwards and spitting. He gave none of them up and, very helpfully, showed his cronies how to spit. He must also have had a special relationship with women because given the chance he would cozy up to women, but not to men, and then bite. Jaja's life at a religious order, where she joined in prayers, left her communicating with gestures, walking upright, and treating other orangutans as beasts unworthy of her company. Some of these habits were so ingrained that they would be extremely difficult to reverse.

Pampering created a different set of headaches. Pampered orangutans tended to be too nice and too naive to tough it out with other orangutans. They were so comfortable with human life that they had little interest in trying orangutan ways. Daidai, the piano-player from Taiwan, was a classic example. Most orangutans gobbled mealworms like candy but when we offered them to Daidai, she got a disgusted look on her face

Juvenile orangutans worked for more than a month to make a hole in their cage floor. They used pieces of concrete scavenged from the sidewalk, then took turns chipping away.

and flicked the horrid things away with the tips of her fingers. Another was Boyke, a beautiful doll whose stunning red hair was brushed lovingly every day; his "owner" had even supplied written instructions for Boyke's care.

Socialization could create new problems, especially when playthings were unavailable. Some, like Otong, spent their time harassing one another. Otong was put in the juvenile cage while she awaited re-release. Thug that she could be, she captured smaller males as her boy toys and bit her way to the top of the female ranks. Hello, one of the most robust orangutans of the group when she was first put into the juvenile cage, was so cowed by the others that she wouldn't compete with them, even for food. She lost a shocking amount of weight and, weakened, became a target for gratuitous attacks. Haishan, her one friend, consoled her after the attacks. Some, bored, resorted to pacing. Seemingly going somewhere with a purpose, they were really just going around and around the cage. I once watched Oscar rotate thirty-six times inside a little niche before he stopped. Ludah gently swung from side to side in the rafters, like a pendulum, for half an hour at a time.

Enrichment

What these orangutans really needed, forest fruits and termite nests, tangled lianas and swaying trees, just couldn't be brought in. Bornean forests support only one to three orangutans per square kilometer, so all the nearby forests would have to have been stripped bare to collect enough to satisfy all the orangutans at Wanariset: not to speak of the labor involved in locating such foods, harvesting them, and transporting them in large volume. The best I could do was simulate the kinds of problems they'd face in real forest situations—finding desirable items hidden *in*, *under*, *behind*, or *through* barriers, and having to calculate travel routes through an interconnected maze of rope or chain "lianas." I was under no illusion that this constituted an adequate substitute for real forest problems, or even sufficient preparation for forest life, but it was all I could contribute at the time.

Little was possible in quarantine, with all its restrictions. But little is more than nothing. We gave the tiniest infants soft cushions and towels to cling to. We brought tidbits of different foods to vary otherwise monotonous days. Once I gave Romeo a coconut before realizing it was too big to pass through his cage's bars. I requested it back, meaning to break it open for him. He refused to hand it over. He negotiated that coconut around his cage himself to a place where he could bang it open. It took him forty-five minutes and it was all solid, concentrated work.

In the socialization cages, our first installations were large oil drums, with both ends cut out; the technicians suspended them by running a chain horizontally through the drum. The infant group swarmed theirs even before it was fixed in place and occupied it continuously for six full days. They were in it, on it, under it, through it, pushing it, swinging it, and falling off it. At night a pile of them slept inside, usually six or seven. The juveniles turned theirs into a cable car. Sitting inside it, they'd grab the chain just beyond the drum and pull, dragging the drum along the chain.

After that, we installed ropes, new climbing structures, and varieties of smaller play objects. Most of what I found was plastic or metal—garish and unsound, ecologically speaking, but cleanable, durable, available, and cheap. We put boat-shaped plastic floats in the older infants' cage, the kind that children use to learn to swim. One little guy toiled laboriously to pull his float up the sloping floor of his cage. Once at the top, he grabbed a cage bar, climbed onto the float, let go of the bar, and slid down the slope like a maniac. Plastic basins were even better. You could slide in them too, plus wear them as hats, sit in them, sit under them, carry things in them, and sleep in them. Every day at sunset, I could enjoy the sight of three or four sleeping infant bodies piled together in one washbasin nest.

Kinoi, a male about five years old, made rings whenever he could obtain a length of hose, then passed himself back and forth through them.

The juveniles' favorite was lengths of garden hose about a meter long. They wove with them, putting the hose in and out through the bars of their cage. They turned them into handles for dangling and twirling, by threading them around cage bars then hanging from the two ends, held together. One, Kinoi, when he was in possession of a hose, invested every second in making giant hoops, carefully inserting one end of his hose into the other and jamming it in tight. Once he'd made his hoop, he passed various parts of himself back and forth through it—an arm, his head, his feet, his whole torso—as if completely fascinated with the idea of going "through" the hole.

I had little chance to see whether these playthings had a major effect on activity levels or learning before it was time to return to Canada. I did have time to see one change: the toys changed the social atmosphere. Ludah, who had previously spent hours on end swinging like a pendulum, became an extrovert once toys were available. Haishan, earlier shy of everything, stuck up for himself when possession of his hose was at risk, even against bullies. I left at least encouraged that I'd brightened a few orangutan days.

Otong creates a shower for herself by standing under the water that technicians are hosing in to wash the cage.

TROPICAL RAIN FOREST HOMES

My real aim was to study forest orangutans, and that meant more than just following them through the forest. Orangutans must understand the forest to survive in it, so I had to acquire some of that understanding too—a formidable task, because orangutans inhabit tropical rainforests. Rainforests in the tropics are among the most diverse and complex living environments on earth. Those of the far east, Borneo and Sumatra included, may be the most complex of all.

For a hint of that diversity, consider this: Borneo and Sumatra represent only 1.3 percent of Indonesia's land mass but they support 10 percent of its known plant species, 12.5 percent of its mammals, and 17 percent of its other vertebrates. Borneo alone has 10,000 to 15,000 species of flowering plants. That is as rich as the whole of Africa, which is forty times as large, and 10 times as rich as the British Isles. In addition, Borneo has 3,000 species of trees, 2,000 orchids, and 1,000 ferns. One tiny 1.12-hectare Bornean rainforest plot included 264 tree species, and that did not include its palms, lianas, orchids, ferns and other vegetation. Borneo's animal life is no less diverse. It is known to support about 222 mammals, 420 birds, 166 serpents, 100 amphibians, and 394 freshwater fish, not to mention the invertebrates, by far the most numerous animal species in

Opposite page: Tropical rainforests are home to an incredible richness of plant and animal life.

One of Borneo's gibbons, Hylobates agilis, shares the forests with orangutans.

tropical rainforests. Many of these life forms are endemic, or unique to the island.

Beyond the diversity, tropical rainforest life flows to the tune of very complex processes that can make surviving a problem of the most capricious kind. Tiny species like butterflies or dung beetles may be able to keep it simple, carving out a niche in a tiny patch of forest and interacting with few other species. But megafauna like orangutans, whose large size forces them to use broad expanses of forest, face a much more difficult task. Most of the difficulties concern food and they probably stress orangutans' minds as well as their bodies. There are three main contributing factors—sun, seasons, and interdependencies.

Reaching for the sun

Sunshine is essential to all plant life; plants make sugars out of carbon and water, and the sun supplies the energy for this process. Tropical rainforests receive twice as much sunshine as temperate forests do over the course of a year. Most of it is trapped in the canopy and never reaches the ground, so plant growth is rich in the canopy but sparse near the light-impoverished forest floor. Plant growth represents food for many species, and popular plant foods, like young leaves and ripening fruits, are most plentiful in the middle canopy and the treetops. So that is where most of the larger mammals and fruit-eating birds feed. Many have even taken to living there and show adaptations for arboreal life. In Borneo, some 45 percent of nonflying mammals are arboreal. Their numbers include orangutans.

Orangutans' enormous size makes arboreality life-threatening. Danny Povinelli and John Cant analyzed the forest structure to suggest why. The rainforest is a vertical world. Tree trunks provide the forest with a relatively sturdy and stable vertical structure for climbing animals, but trunks are separated from each other horizontally, so getting across the gap can be a problem. Branches sometimes offer a bridge between trunks, but they taper towards their ends so the farther they reach, the weaker and less stable they become. They also bend under weight, so they may not take climbers where they aim to go, and they may even break. Lianas sometimes offer an alternative bridge but they too vary greatly in strength and stability. The heavier the climber, the worse the problem. Good supports become fewer, gaps become effectively wider, branch bending worsens, and the cost of falling may be serious injury or even death. Orangutans, the largest primarily arboreal mammal in the world, must be pushing arboreal supports to their limits. There is ample proof. Their skeletons often have broken bones, they quite commonly fall, and they steadfastly refuse to jump.

These arboreal difficulties must make great demands on orangutan minds. Orangutans

Bandang, Borassodendron borneensis, an endemic palm, is an important source of food for orangutans in times of fruit scarcity.

In climbing towards the sun, species without the strength to climb on their own climb on other species. Strangling figs start their climb as slender vines, then finally grow so large that they literally strangle their supporting tree to death.

have devised special techniques for travelling arboreally that appear to entail very high-level calculation of physical possibilities. For instance, in tree swaying they climb from a large tree onto a slender, pole-like "vehicle" tree then pump it much as one would pump a swing, so that it sways toward the next large tree they wish to enter. They sometimes pull on adjacent vegetation to aim their swaying in the right direction. When their vehicle tree swings close enough to their target tree, they grab its leaves, pull themselves closer, and climb on. In clambering, they pick their way through the tangle of lianas and branches. To do this, they must determine what combination of branches and lianas can span a gap, how to distribute their weight across multiple supports to forestall breaks, and how to test and correct for bending and cracking on the spot. Povinelli and Cant argue that clambering requires mental abilities as complex as any seen in great apes.

The seasons

The "rain" part of the name rainforest refers to the fact that these forests receive large amounts of rain every month of the year. Rain paces the seasons in the tropics so Bornean and Sumatran rainforests have wet and dry seasons, not hot and cold ones. These wet and dry seasons affect *when* particular foods are available as well as *where*. But this seasonality is weak, and whimsical rather than regular. The area around Wanariset, for instance, recorded dry seasons in 1994 and 1997 but not in 1995 or 1996, and the dry season ran for three to four months in 1994, starting in August, but almost for eight months in 1997, starting in July. In the years I visited Camp Leakey in south-central Borneo, the dry season ran from June through September or October, that is, months earlier than at Wanariset. If there is a prevailing seasonal pattern, it is *microclimates*, which affect very small areas or particular habitats. Even the climate within the canopy differs from that which exists above or below it.

Borneo's hundreds of species of rattan, the spiny climbing palm of cane furniture fame, make Borneo among the world's most important centers of rattan trade. Rattans are among the many forms of plant life that climb high in the canopy to find the sun. Orangutans eat rattan fruits and hearts. This orangutan is reaching into the rattan.

When the season is right, the forest is filled with butterflies.

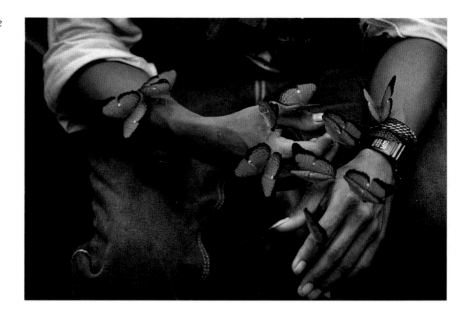

These seasonal patterns can make conditions so fickle and localized that predicting when and where fruit will be available must involve complex understanding. For fruit-eating species like orangutans, that turns finding food into a serious problem. The mix of conditions that triggers fruiting in any one species can be so ephemeral that fruiting could occur twice in one year then not again for several more, and so local that one tree could have ripe fruit while a second, a mere 100 meters away, does not—it may not fruit at all that year, it may have fruited weeks earlier, or it may fruit weeks later. One species' fruiting patterns may tell you nothing about others' because each species may react to different cues. Plants seem to live out of synch with everything around them, even their own species.

Even the lack of synchrony is unreliable. Every few years forces conspire to synchronize plant cycles. Many species then fruit together, a phenomenon called mast fruiting, climaxing in enormous riches. What appears to trigger masting is prolonged drought. For fruit-eaters the final feast is great, but it may be bracketed by famine-like conditions. Massive fruiting has occurred twice since I've visited Wanariset, in 1994 and 1997, following their severe dry seasons. The intervening years, 1995 and 1996, with virtually no dry seasons, produced very little fruit. Masting was predicted in 1997 because it often occurs in El Niño years, but the unparalleled length of the drought that year left many species unable to bring their fruit to maturity.

Interdependencies

Interdependency among life forms, a signature feature of tropical rainforests, adds yet more layers of complexity. Tropical rainforests are abundant with epiphytes, plants like orchids and ferns that fasten onto other plants for support, lianas and rattans that climb the trunks of sturdy trees to reach light high in the canopy, and myrmecophilous plants that have symbiotic relationships with ants.

With everything living off everything else, many plants become prey when they'd rather stay alive. Probably for this reason, tropical rainforest plants are renowned for their defenses, devices designed to protect the plants' vital resources from predators.

Many species defend their resources by hiding them. Some put them out of sight: inside shells, underground, and beneath bark. Some put them in inaccessible locations. The rattan's heart, its center of growth, almost floats in midair at the tip of the cane held up only by tiny grapnels along the rattan's leaves and stem that snag other foliage. Some hide them in time, like plants that cycle out of synchrony with one another. And others hide them by dispersing—that is, they space their individual plants far apart rather than clumping them closely. If one individual of the species falls prey to a predator, at least it does not guide the predator to others.

Other species fight back. Some plants use physical defenses like the sharp spines of durians or the rock-hard shells of coconuts. Others produce noxious chemicals, many repulsive to the taste and some outright toxic, like many species in the mango family. One mango family species, *renghas*, exudes a toxic sap so vicious that one drop burns human skin black and rots it off. Defense chemicals are prevalent in forests on nutrient-poor soils where it is costly for trees to replace leaves eaten by herbivores—like the forests of Borneo.

This all affects orangutans. The main reason is that they are too big to survive on fruit alone, like other great apes. An orangutan can eat a tree's entire fruit crop in a few hours, yet come nowhere near satisfying its daily nutrient needs, and fruit provides little of the protein and fat needed for health. So orangutans meet their nutritional needs by diversifying their diets to a wide range of food types. That alone strains mental capacities: it can entail recognizing up to 400 different foods and knowing how to find each.

Probably worse is *what* orangutans eat beyond fruit. Fruits may be hard to find but they are usually easy to handle. Fruit has been described as the only form of life that "wants" to be eaten and it is often designed to attract predators—colored for best visibility, shaped and sized for easy manipulation, and relatively free of defenses. The foods orangutans eat to supplement fruit, however, like leaves, shoots, and colonial nest-building insects, are a different kettle of fish. Those rich in fats and proteins tend to be vital to the host organism's life—its own energy stores, reproductive material, or growing center. So many of the supplementary foods are difficult to handle because

Borneo supports 28 species of nepenthes, carnivorous pitcher plants. Their chalices, exquisite to look at, are designed to trap, drown and digest insects. Lured to the brim of the deadly pitcher, the insect falls to the watery death waiting below–unless other lurking predators snatch it first.

Among the forms of life found on the forest floor, fungi live off dead plant matter and in some cases transform it into edible food for other species.

they are the very items protected by anti-predator defenses. Some of these supplementary foods cannot be avoided, difficult though they are to obtain, because they serve as the few "fall-back" or "permanent" foods available when fruit is scarce.

That forces orangutans to face a profusion of defenses. Termites protect themselves within the cement-hardened walls of their nests. Rattans protect their vital growing hearts beneath leathery sheaths covered with millions of needle-sharp spines. Many leaves and ants produce chemicals toxic to would-be consumers. For orangutans, this translates into having to devise intricate strategies and complex techniques for foiling a plethora of nasty defenses.

Other great apes face similar problems and they are known to obtain difficult foods using techniques that demonstrate high intelligence. West African chimpanzees use stone tools to crack hard nuts—not just single stone hammers, but sets of stones serving as matched hammers and anvils. Tetsuro Matsuzawa, who studies them, even found chimpanzees adding a stone wedge to level uneven stone anvils. Not only does this add a third tool to the set, the wedge tool modifies the other tools. That implies minds that can improve their own inventions: in mental terms, reconsider and modify their own ideas—or reason. Dick Byrne, who studies mountain gorillas, found the gorillas had invented a slick way of eating stinging nettles, a complex system of folding and wrapping that hides the stinging hairs away from the mouth. That technique too is complex and shows the same high-level intelligence as chimpanzees' tool-using.

As for orangutans, they have some physical adaptations to beat anti-predator defenses. They have a high tolerance for foul-tasting chemicals, for instance, and are simply not repelled. They have massive jaws that can crumble termite nests or coconut shells. But for some foods they too use fancy techniques. They chew into small bits some foods that are astringent to our tongues, probably full of tannins, then spit-and-suck or rub the mash through their forearm hair before consuming it. This creates what looks like soap lather. However it works, the lathering may some-

Even richer than the array of pitcher plants is the wealth of orchids, one of the rainforest's many epiphytes.

how alter the chemistry of the food to render it more palatable. Another ploy they may use against chemical defenses is eating clay. Clay can act to buffer some of the toxins, probably polyphenols and tannins. Perhaps most convincing of all, it was recently discovered that orangutans do use tools for food. Sri Suci Utami, an Indonesian researcher, observed a wild Sumatran female make leaf "gloves" and "cushions" to protect herself from a particularly spiny food. Carel van Schaik, who studies wild orangutans in an unusually swampy part of northern Sumatra, recorded them making and using a variety of tools to obtain inaccessible foods, mostly probes, hammers, and levers. For orangutans too, food problems make hard mental work.

Arboreality reconsidered

Arboreal life worsens orangutans' feeding problems. Fruit, for instance, is often located on flimsy terminal branches that are nowhere near strong enough to support an orangutan's weight, making access to those fruits problematic. Many of the non-fruit foods are located in the tree-tops too, so orangutans must negotiate supporting themselves in the trees at the same time as they apply complex techniques to the difficult task of handling the food. Unfortunately, the two tasks complicate each other. Exerting force on the food may disturb the orangutan's precariously maintained balance, for instance. Anyone who has ever fallen off a ladder while trying to hammer a nail has a taste of what orangutans must experience. The kind of intelligence that can coordinate all this must be able to handle not just each problem independently—each one a complex mental feat on its own—but the bigger task of the two problems intertwined.

Acquiring the expertise

The expertise orangutans need to cope with the complexities of life comes very little from "instinct" and very much from their own learning experiences; in this they are like other primates. They are lightning learners with vast memories, and they may be capable of exceptionally sophisticated understanding and insightful thinking, although compared with humans, their learning is slow and their understanding rudimentary. It takes orangutans, like other great apes, eight to ten years to develop the knowledge, skill, and understanding to survive independently in the forest. Furthermore, the knowledge and expertise normally available to other primates through their social group is much less accessible to orangutans. As dependent infants, their only reliable social companion may be their mother; as semi-independent juveniles and adolescents, their social contacts are relatively infrequent; and as adults they live primarily solitary lives. For orangutans, then, acquiring the expertise to dine on forest fare and travel effectively through the trees is an extremely challenging intellectual task. This process of acquiring forest expertise was what I hoped to study closely, in the orangutans of Sungai Wain Forest.

LORD OF THE FLIES

I first ventured into Sungai Wain Forest in 1995, in search of opportunities to observe forest orangutans' minds in action. On this trip, the first of what turned out to be annual expeditions, I opted to concentrate on their foraging skills. This is the area in which orangutans, like other great apes, are considered to face the greatest challenges to their intellect.

Awaiting me was an orangutan *Lord of the Flies*. ORP had released about 60 ex-captive orangutans into Sungai Wain, mostly juveniles. Because the forest had no wild orangutans, these orangutan children were running their own world. It was a very abnormal society. Charlie, the infamous tough guy of the Taiwan Ten, but still only a nine-year-old adolescent, lorded over all with the swagger and circumstance of a child king. With no older orangutans to take him down a peg, his manners were no nicer than they had been in Jakarta. He had appeared at Daidai's release, for instance, to check the newcomers and get them into line. He kidnapped Daidai immediately, and literally dragged her everywhere with him. Helga Peters, who was studying the newly released orangutans, said Charlie wouldn't even let Daidai go to the feeding site for provisions. Daidai knew nothing of forest foods, so she starved. She tried to escape. Once, for instance, she tried to sneak off while Charlie was wrestling

Post Sinaga was one small, lonely post in the forest.

another male, but he spotted her move, shot out one hand, and clamped her back in place—all without missing a beat in his wrestling. Daidai escaped after a month, but she was wounded and so emaciated that she was returned to Wanariset.

I made my way into the forest in June with my carload of equipment, supplies, and helpers. We rendezvoused in Sungai Wain village at the edge of the forest with Dolin, the ORP technician who had agreed to assist me in the forest. There we were also met by one of Sungai Wain's orangutans. While we were loading up, Dolin took me behind a house to a small cage filled with Enggong, a male about five years old, who had recently left the forest and entered the village. He was stashed there until staff could carry him back into the forest—where he was expected to stay.

We left Enggong waiting patiently and started out on our trek. The terrain was uninviting. We crossed a swamp reputedly infested with crocodiles, then hiked to a lonely, empty post. A cable car bridged the swamp for the non-hardy. It consisted of an angle-iron box that dangled from a cable run between platforms at either side of the swamp, with a motor to crank it back and forth. It was a tossup which of these two evils—wading through swamp or braving the cable car—was the lesser. With all my supplies, I opted for the cable car. After its first lurch had bashed my forehead against the angle-irons, it quickly picked up speed, whizzed through feathery rattans and bamboos, bogged for a while in mid-swamp under our weight, then whizzed again as we entered the home stretch. At that point, I discovered we had no brakes. To stop, you shrieked to the operator to cut the motor. Our shriek triggered a final, violent lurch that stopped just shy of catapulting us off the end of the line.

Past the swamp, the trek to the post was a mere half-hour walk along a well-traveled trail. Even so, the equatorial humidity and heat made it feel like climbing Kilimanjaro. We entered a world of palms and rattans, streams and hills, very unlike the environment I'd known in Tanjung Puting. Up one of those hills we came to an opening in the forest and there was Post Sinaga. Humbler than Camp Leakey, it had one square room, a patched-on kitchen-bathroom, and a narrow raised porch in between.

It was less lonely than its advance billing had promised. In fact, there was a welcoming committee: Charlie. I knew him instantly, even four years after I'd last seen him in Jakarta. How he knew someone would be there when the post had been empty for two months isn't clear. My guess was that the cable car clatter and our retinue filing in alerted him. He fit his reputation: face tense, brows knit. I wasn't too worried. He was, in my mind, just a kid—until he raced straight at me with a face that made it clear he was dead serious. I moved out of his way, fast. He grabbed past me for a bag of soap flakes, got it, and split.

Eventually our helpers departed too, leaving Dolin and me hanging mosquito nets over beds, bringing water from the stream and boiling some for drinking. We cached our goods in humidity- and insect-proof containers, and cleaned away debris left by the insects and mice who had claimed the post while humans were absent. Night fell promptly around six, on went a kerosene lantern, a kerosene stove, a radio for talking with the village, and flashlights. By nine I was ready for bed. I fell asleep to strange sounds, but I woke at the next day to familiar ones: a dawn chorus of gibbons.

The players

Next day we headed about two kilometers north of the post to the forest's third release site, *kandang* 3 or K3, meaning cage 3. Orangutans had been released there in February 1994 and March 1995 and several regulars still came for daily provisions. The "kandang" name became clear as we arrived. A giant cage loomed above a forest ridge, like a rough-cut wooden boxcar mounted on six-meter stilts. I knew how it was used. At a "release," humans carried the ex-captives to be released to the site in small cages, deposited them all inside the giant cage, and locked the door. The group was left there for several days and fed daily, theoretically to calm them down. A formal ceremony celebrated their true moment of freedom. On the chosen day, the trap door was opened before a crowd of volunteers, officials, and media, witnesses to the orangutans' cautious exit to freedom and their first blinking looks at their new world. Thereafter the cage became a giant picnic table, where daily provisions of bananas were served around 9:00 a.m.

Semoi climbs onto Dolin's back.

Two little male regulars were at K3 for provisions on my first day. Semoi and Paul, both about five years old, were newcomers who had been in the forest only four or five months. Semoi was especially amiable, and adored forest life, but he seemed lonely, probably because he was barely past infancy. Only days after I arrived he walked out of the forest into the village. Staff returned him to the forest, to a site farther from humans, but he marched right back out. We had little choice but to return him to Wanariset to grow up a bit. Paul, more retiring, had had mishaps too. When first released he was somehow wounded and had to be returned to the clinic. He recovered and was re-released at K3, but K3 and its orangutans were unfamiliar to him and he was not faring well. His ribs showed, his hair was ratty, he had several behavioral tics, and he didn't play. He looked as if he was trying to cope, and we didn't want to banish him to the clinic again, so we decided to monitor him closely to diagnose what was wrong.

Enggong was soon brought to K3 and he turned out to be gentle-tempered, independent, and very busy. He'd been in the forest over a year and had no blots on his record, barring that village sortie. His success may have been due to his past life at Bontang, the same old-style rehabilitation project Panjul came from, where he had racked up many forest hours.

It was a couple of weeks before I met Bento, the fourth K3 regular. Late one day Dolin warned that he was near. Bento was about six years old, quite big, a two-year veteran of Sungai Wain, and a troublemaker. I had to face him if I was to work with him, so I squatted down and waited for whatever might happen. Bento approached me boldly, stopped an arm's length away, and looked me in the eye. I looked back. After a minute he reached out one finger, touched my right eye, picked at it, then sniffed his finger. My mind registered "Cream-puff." Bento walked on. That was it. I had not one stitch of trouble with him, other than his attempts to grab me to play.

Sariyem was the only female who visited K3, and she came only from time to time. She was forest savvy too, a third orangutan transferred from Bontang. When she arrived all the boys went mad with showing off and squabbling about who could play with her. Bento and Enggong, the bigger and stronger males, had the edge in putting on a show but Sariyem passed them both over for little sober-sided Paul. She operated very differently from the males—she worked and worked, slow and meticulous, with none of their fooling around. It could have been that she was alone, or, just that she was like Paul—a very serious orangutan. I suspect that at least some of the difference was because she was female.

Charlie's personal headquarters were to the south, but periodically he

stormed up to K3, the same way he visited Post Sinaga—tense, glowering, and pugnacious. Charlie at K3 was on patrol, checking who was where and keeping them all in line. After he commandeered food (which was easy, because everyone backed right off), he "invited" the males to "play" with the same politeness he had offered Daidai: he dragged them down the hill and wrestled so roughly I winced. Orangutan rules apparently forbade running away because they just took it, cringing. K3 didn't relax until Charlie's storm went south again.

There were also orangutans from early releases—Tuti and Uce, females, and Aming and Imelda, a shadowy male-female pair. They occasionally appeared in southern areas of the forest, but never as far north as K3. I hoped to bump into them sometime later.

Roughing it in the bush

The daily routine I came to know as typical of the K3 orangutan boys' club wasn't far from the Camp Leakey one: arise near dawn, forage a while, head to K3 in time for provisions, rest and play until late morning, then amble off into the nearby forest to forage some more. Sometime between five and six in the afternoon, they would head up, sometimes way up, to nest. Below, we'd eventually hear the sounds that released us, too, from our day's work, the

Sariyem is furious because Bento has just stolen her food.

Bento approaches the ground.

hearty cracking of strong branches that signalled nest-building.

What I could see of their travel patterns was that they didn't wander far from K3, at most a couple of hundred meters, and they sometimes circled back to nest near it. So like the Camp Leakey rehabilitants and wild orangutans, they followed the principle of nesting close to tomorrow's breakfast. Also like other rehabilitants, they traveled a lot on the ground and they made use of human trails in navigating. They all negotiated arboreal movement with finesse—I saw no klutzes as bad as Supinah, for instance. And they slept in nests more often than other contraptions, although a few still liked the top of K3 or Post Sinaga's roof and they occasionally reused old nests.

The K3 crew were still youngsters, which probably accounted for the obligatory play session, the very small traveling range, and the outrageously high nests. Other than that, they didn't lurk about the Post for food at the end of the day or mooch from humans, because it was impossible in Sungai Wain. I could even leave my backpack on the ground without fear of losing it.

Semoi, Enggong, and Paul gave me my first look at how they ate. They had ambled through a great tangled thicket of lianas and barbed rattans, so I fell behind and lost sight of them up a hill. When I finally hauled myself to the top, all three were sitting in a circle at a termite nest—dirty faces, pot-belly tummies, floppy duck-feet sticking out in front of them, breaking chunks of nest and sucking termites from their exposed cells.

The food's the thing

Paul (on top) and Enggong (laughing) indulge in hearty wrestling.

It was indeed the nasty foods, the defended ones, that presented the hardest problems. The orangutans faced rock-hard coconut shells, termites nests, ironwood bark, and stings that left dizzying pain. Nutritious rattan hearts and rattan fruit grew at the very tips of rattan canes, suspended fifty meters high in the canopy. Other foods hid high in the crowns of towering palms, some sporting an army of protecting leafstalks.

They had some unusual operations for handling these problems, which perhaps are unique to orangutans. They used lathering, which I initially thought was a bizarre Camp Leakey ritual for eating soap, to eat some *Diospyros* fruits and some leaves. They had a rich kit of operations for bark. They removed long strips of bark by first "scoring" it, biting repeatedly along a length to serrate its margins. Then they bit hold at one end and with one prying jerk, tore the whole strip off. On one species, they nibbled bark off in a pattern to make long trenches lengthwise along the undersides of branches. It was no fluke. On one of these trees, almost every single branch had one of these trenches scarring its underside. They even had a variety of special operations for eating termites. They mostly sucked them from their nests, but when Paul found an especially rich nest, he simply held it over his head and tipped it like a cup. In alarm, thousands of termites swarmed to the surface of the cells and tumbled off, right into his waiting mouth below.

These operations were mere details, compared to the complete procedures for handling nasty foods. I had to wait until 1996 and 1997 to see them, when the ex-captives had advanced their skills and I had a better understanding of what they faced.

Heart of palm was the most difficult food I saw them tackle. Their favorite was the heart of a tree palm, bandang, *Borassodendron borneensis*. The heart is the palm's only center of growth—if you remove it, the palm dies—so it is fiercely defended. It hides deep within the crown, defended by dozens of massive leafstalks with edges like razors. Many forest orangutans' hands are scarred from those edges. Orangutans obtain bandang hearts by pulling out a palm's newest leaf just as it emerges from the center of the crown, like a spear. The tender tissue at the spear's base is heart material, which they simply bite off and eat. The job is more complicated than it sounds. Mature bandang grow ten to fifteen meters tall, so their crowns hover high in the air, and their new leaf spear can be as thick

Paul eats a bandang leaf stalk. This operation involves pulling the stalk apart while dangling from an adjacent tree.

around as a woman's wrist, and so sturdy that only adult male orang-utans have the strength to extract it.

My juveniles had invented an elaborate procedure to circumvent the strength problem. They subdivided the spear into several smaller sections that were easy to pull out. They could do this because bandang leaves are giant fans on long stalks, each fan composed of fifty to sixty small blades, or laminae. As a new leaf grows, the fan emerges, tightly

Enggong tries to get food from a bandang in the midst of its crown.

closed, in a long spear. Simply picking at the spear's tip easily separates individual laminae. So the juveniles fingered off small sections of a few laminae each then pulled each section one by one. They often had to separate the blades into five or ten sections, and it could take as long as twenty minutes to get the whole spear, but the technique worked.

There was more. Before starting on a spear, they prepared the crown for work. They meticulously removed debris, pushed older leaves obstructing the spear out of the way, and even made a nice place to sit. They chose an older leaf, then either pushed it down to a horizontal position and sat at its base or bent it over the spear and sat on top. Only when things were properly set up did they start work on the spear itself. In addition, they would often bend a section's tip and bite over the bend before pulling. Doubling over probably strengthened the section (tips tended to snap easily). They also ate the heart in courses, the way you'd eat asparagus. They ate the best part of each section first, then set its remains to the side. After they'd eaten the best of every section, they gathered up the remains and moved to a comfortable spot, where they finished off the less delectable parts.

When Paul found an especially rich termite nest, he came up with this trick of drinking them.

143

Paul pulls the new leaf from a small bandang palm for the rich and delicate heart material at its base. He isn't yet able to handle mature palms, only those accessible from the ground. He uses an adjacent tree as a brace to strengthen his pull.

Handling termites also involved complex procedures. The nest-building termites eaten in Sungai Wain do not build the great mounds that are famous in Africa; their nests are smaller and shaped like basketballs, lumpy footballs, or drooping lobes. So the orangutans don't probe into termite nests the way chimpanzees do; instead, they break off a chunk of a nest, crack it open, then suck termites from its exposed cells. By 1997, Paul had developed the most elaborate version of this procedure.

Paul first collected several chunks of termite nest, not just one, tested each for termites, and discarded any that were empty. Once he had enough (usually two or three chunks), he took one chunk in one hand—the working chunk—and held the others in his feet to be eaten later. Then he set about breaking the working chunk open. He cracked off a small piece with his hands and teeth, lifted it off like a lid with lips or fingers, sucked the exposed surface to extract the termites, and ate them.

If the piece was rather big, Paul might break smaller bits from it and suck the bits. If the bits were big, he might break crumbs from the bits, then rubble from the crumbs. He even created a system for remembering what was what. He held the working chunk in one hand and all his fragments in the other—the piece between thumb and forefinger, the bit pinned at the base of his palm with the tip of his third finger, a crumb balanced on his wrist, and any rubble balanced farther up his forearm. Then he ate all his bits and pieces in order, from smallest to largest: As he cleaned out each piece, he discarded it and went back to the piece he had broken it from. When he had finished his whole chunk, he moved one of the chunks from his feet into his hand and started all over again. The job could take over an hour.

If this seems like excessive talk about food, remember that everything reminds orangutans of food. Their health, their encounters with humans, and their own social lives revolve around food. So, it turns out, do many of their abilities, including tools, imitation, plans, and deception.

What's eating Paul

Paul's problem in 1995—weight loss, poor hair, behavioral tics—boiled down to food. He foraged in the same area as Enggong and Bento, but they were more forest-experienced and they were buddies, so they usually took the lead and foraged together. It would have helped Paul to tag along but they wouldn't let him.

Paul was finding food on his own though, and when he was alone he behaved like a normal and hard-working, if serious, little orangutan. His problem was Bento. Bento turned out to be a food thief. He regularly stole the food Paul found, so Paul's hard work was going into Bento's fat stomach. Paul tried many tactics to foil Bento, from hiding his food to running or sneaking away with it, but Bento dogged him, sometimes all day. Not only that, Paul *would* eat slowly, especially the banana provisions. If he had just stuffed them into his mouth and run he might have kept them, but his genteel nibbling was an engraved invitation for Bento. It was no wonder Paul was losing ground.

So we devised a protection scheme, guarding Paul's eating when Bento was near and passing him food out of Bento's sight. We also gave him extras, like eggs and vitamins. I left for Canada just as our system got under way. When I returned seven months later, Paul was much improved: his hair was better, his tics were fading, he had a nice little pot on his stomach, and he could cope on his own. Food fixed, problem solved.

Encounters

Our encounters with orangutans were almost always at their discretion, and also invariably, had to do with food. We probably wouldn't even have met the K3 regulars if it hadn't been for the provisions we brought. I finally met Aming and Imelda through food, although it was only because they made a mistake. Aming was a male about eight years old, and Imelda a female a year or so younger. She was another of the Taiwan Ten, and another name-mixup victim. Manis in Jakarta, she became Imelda in Wanariset. She and Aming had been a close pair since their release together in 1992.

Dolin and I had just returned to Post Sinaga after a follow. All was peaceful as I peeled off my leechy socks and savored the thought of the

Paul's behavior–sitting out in the open, savoring each bite of his banana–is an open invitation for Bento to steal his food.

cool water Dolin was getting from the kitchen. But instead of bringing water, Dolin shouted: "*Wah-duh ada orangutan di dalam!*" Roughly translated, that's, "Yikes, there are orangutans in here!"

In our kitchen, huddled high on a shelf, were Aming and Imelda. They must have come in by squeezing through the bottom corner of the door. Once in they couldn't pull it enough to get back out, so they were stuck until we came with the key. The kitchen was the disaster area typical of orangutan break-ins. Anything they deemed food had been eaten: a kilo of sugar, a wok of fried rice, two boxes of cookies, three chocolate bars, a can of condensed milk, and five packages of noodles. If we were aghast, they looked terrified, about as brave as a couple of homeless urchins caught with their hands in a cookie jar. Aming tried to be brave and protect Imelda but he was so scared he squeaked. We couldn't even shoo them out. Being orangutans, they just retreated higher and higher if we thrashed at them. We had to leave before they got up the courage to flee.

The next day I went to town to replace the goods, and also the locks and hinges. When I got back they'd done it again, seizing their chance when Dolin left the Post for ten minutes to bring water from the stream. With impeccable orangutan timing, they took advantage of humans' predictably short attention span. On day three Dolin returned to the Post to find Aming prying the kitchen door open with a pole. Foiled this time, they got bored and vanished.

Tools for food

What Sungai Wain juveniles did with their food made me take a second look at tools. Technically, if you recall, tools have to be *detached* objects, but for hulking arboreal creatures like orangutans, "detached" may amount to a four-letter word: detached equals falling and falling equals injury, if not death. Orangutans are almost *never* completely detached

Aming looks intently at the camera.

from support. And there are certainly objects that help attain a goal *because* they're attached, like ropes to climb trees. So a defining criterion for "tool" that makes neat technical sense may make biological or functional nonsense—in which case, it is worth considering what happens if you drop that technicality.

If you drop the "detached" criterion, it's clear that orangutans use numerous tools, some for obtaining or eating food and some for negotiating the trees. They use pitcher plant chalices as cups, drinking what is probably a "soup" of partially digested invertebrates inside. They bunch up leaves to make wipers to clean their bodies. They rarely detach the chalices or the leaves. Why bother? They work the same way attached as detached. They use slender trees as braces to balance forces when they pull foods, modify palm leafstalks then sit on them as workbenches, and bunch branches and lianas together to make handles to support their weight while they work in the trees. Enggong once broke a branch to make a tool, but deliberately left it attached to the tree. He cracked it so that it was loose but not detached, and held the dangling end like a handle. That gave him the few extra centimeters' reach he needed to enter a palm.

Food for imitation

Over the years, I watched the K3 boys improve their food procedures, and their learning appeared to involve imitation. In 1995 neither Enggong, Bento, nor Paul could obtain bandang hearts. They ate them, but only if Sariyem came around, because she knew how to get them and they knew how to scrounge from her. By the time I returned in 1996, seven months later, all three boys could get bandang hearts on their own.

Enggong used something suspiciously like imitation to learn about a new food from Aming. Aming and Imelda were transferred to K3 in March, 1996, after they were caught raiding fruit in Sungai Wain village. They didn't know the K3 area, but they brought their great knowledge of forest life with them. A few days after arriving at K3, Aming entered a big tree near K3 and started eating its bark. The bark must have been extremely tough because Aming's work made loud cracking sounds. Enggong apparently heard the sounds, because within minutes he was in the tree too, dangling close to Aming and watching him intently. Usually, spotting a companion eating inspired the observer to eat too. It was easy with bark—there was no monopoly on the tree. You could just

Enggong watches Paul intently as Paul eats from a termite nest. Enggong will beg for a share and later scrounge Paul's leftovers that fall to the ground. The careful watching can contribute to imitation.

take a position on another branch and chew away. Instead, Enggong gazed fixedly at Aming's work for several minutes, moving only to watch more closely. When Aming finally left, Enggong immediately took the spot he had just vacated and bit at the bark exactly where Aming had been working.

What Enggong did is what I've seen other orangutans do with foods they don't know. He ate *exactly* what he had just seen eaten. There were more clues that Enggong was trying something new. Aming had been able to tear off big pieces of bark but Enggong broke off only small crumbles. Aming worked for an hour, but Enggong gave up after a few minutes. The tree was one that the K3 regulars passed virtually every day, right on the main trail to K3 and no more than thirty meters from K3 itself, yet there were no signs that any of the K3 regulars had ever eaten from it: no scars from prior bark-rippings and no food-tree label. In two years of working in that area almost daily, Dolin had never seen any of the K3 regulars eat it. But Aming recognized this tree as a food source within days of arriving, and knew exactly how to eat from it.

This was the closest I was ever likely to come to a provable case of imitation "in the wild." What other conclusions can be draw from the facts? Aming must already have known how to get bark from this tree, Enggong must not have, and Enggong must have learned about this food and something of how to get it by imitating.

Planning for food

The ex-captives showed the same patterns as those that suggested planning in wild orangutans. Several of them, instead of eating whatever they met along the way, scheduled their foraging in blocks, which had to involve some sort of a plan. Enggong, for instance, would eat nothing but termites for a whole hour or more, passing over bandang hearts and many other foods as he did so, then suddenly shift and eat nothing but bandang hearts for the next two hours, now ignoring all the termites he passed.

Orangutans could concentrate so much on the search for one food that they missed all manner of other foods. Once a technician, Iyan, and I were following Enggong, Bento, and Paul near the K3 trail. We'd moved ahead so we were the first to spot a pile of rice in the center of the trail (another technician had jettisoned it as he fled from Bento). That rice had *Eat me* written all over it—white against the forest, completely exposed. Incredibly, all three orangutans walked right past it, a food they loved. Iyan and I were so astonished that we blurted out the equivalent of "Hey, you guys" and pointed madly at the rice. All three looked up as if stunned at our interruption, confused until they saw what we were pointing at. Then they were on the rice like a shot.

Their travel patterns also suggested plans about food. Near noon one day Enggong headed south along the ridge from K3, along with Bento and Paul. They made a straggly group, weaving together and meandering apart, just snacking on bits of this and that. About 1:30 in the afternoon Enggong disappeared high into a tree. In a few minutes we heard things falling through the foliage then saw fruit seeds hit the ground; we later identified them as *Microcos crassifolia*.

When Enggong came down an hour later, he abruptly changed his travel direction. Before he'd been heading mostly south, now he turned northeast. He passed over hills, across valleys, along a trail, and through a swamp, periodically pausing to eat this and that, but always keeping his northeast bearing. At 5:00 p.m. he disappeared up a tree. Knowing the daily rhythm, I hoped he was heading up for his last feed of the day before nesting. There was silence for a bit; then, sure enough, we heard things dropping. Enggong was eating. I went to see what was falling, and what did I find but the same *Microcos crassifolia* fruit he had been eating at 1:30. He had changed his direction abruptly right after eating from the first *Microcos crassifolia* tree, stuck to the new direction for almost three hours, and when he stopped, he was at another tree of the same species. Can there be any doubt that he planned it? Having discovered ripe fruit in the first tree, he set out with the intent of eating from the other tree as well, which he obviously knew and remembered.

Deceptions

There was even deception in the woods around food. The best case pitted Paul against his nemesis, Bento. Bento had come lurking about while Paul was peacefully eating termites. It was 1996, and Paul was now attuned to Bento's harassment. Paul discreetly shifted to a spot where he was half-hidden from Bento's beady mooching eyes, yet could furtively monitor his movements. Nothing alerted Bento and he continued his own work, taking the heart from a nearby bandang. Paul loved bandang hearts, but he hadn't yet worked out how to get them on his own. He was in the habit of scrounging others' leftovers. But this was Bento. So Paul waited until Bento was turned away. Then he crouched low and sneaked up to Bento's bandang, where he very quietly and carefully picked up some half-eaten spear sections that Bento had discarded, then just as quietly and carefully sneaked them back to his secluded spot. There he turned his back on Bento and ate them, frequently checking over his shoulder for any reaction. He was not detected.

Finally, Bento left the bandang. The rest of his leftovers were coming up for grabs. Instead of lying low until Bento was out of sight, as he usually did, Paul ran directly in front of Bento to K3, twenty-five meters away, then—right in front of Bento's eyes—ate papaya leftovers. The result was entirely predictable: Bento hassled Paul for the papaya and Dolin and I had to fend him off until Bento gave up and departed, leaving Paul to eat his papaya in peace. But instead of staying, Paul sped back to Bento's bandang, climbed into the crown, retrieved the leftovers that had lodged there when Bento had discarded them, and settled in to eat them.

I couldn't convince an inquisition of scientific scoffers that Paul deliberately pulled a fast one on Bento, acting as his own decoy to lure Bento away from food he wanted. But Paul's behavior was out of character, it made use of predictable patterns in Bento and in humans, and it removed a disturbance that regularly interfered with Paul's eating. Putting this sort of thing together with reports of similar strategic action in other great apes, I concluded that Paul did deceive Bento, and he did so in order to filch Bento's food.

Is there life beyond food?

There were exceptions to the rule that food reigned. For males, it was a toss-up whether food or females took top priority. It was a confrontation about a female that gave me my final opportunity to see Charlie, Aming, and Imelda that year. A film crew had stopped to rest at Post Sinaga after their forest trek when two orangutans, Imelda and Charlie, emerged from the forest. Charlie hadn't handcuffed Imelda the way he did Daidai, but he was keeping her close as they moved around the porch—he, tough intruder, she, hunched hostage. While we were enjoying this close look at them Aming arrived, likely in quest of his Imelda. He came so quietly that I didn't notice him until Charlie gave chase, and Aming fled, squeaking in fear. Charlie stormed after

him, and the chase exploded into a swirling, crashing, screaming race through the trees. Charlie caught Aming and grappled with him, biting him until Aming tore free and ran. Not satisfied, Charlie pursued him far into the forest.

So stunned were we that we had to shake ourselves back to normal once the fury of their battle faded. Only then did I notice Imelda, still there, sitting by herself. She too appeared to be just returning to normal, straightening up from her hunched position and looking around the Post attentively. Neither Aming nor Charlie was anywhere in sight. Then an odd expression spread over her face, one I've only seen on orangutans being goofy, and she hopped jauntily off her perch, somersaulted twice, and scampered into the forest with her arms waving over her head, in the opposite direction from Charlie.

Charlie arrives at Post Sinaga with Imelda, whom he presumably kidnapped from Aming. Imelda's hunched posture suggests she's not a willing partner.

RELEASE ME

It wasn't until 1996 that I saw an actual ORP release. I arrived as preparations were gearing up. Orangutans were being chosen for release; the release cage (K6) and a new cushy post (Post Djamaludin) were being built at a site six kilometers into the forest; and more technicians were being hired to support the newly released orangutans. Workers were cutting trails to allow human observers to follow them. Then there were arrangements for a helicopter to lift the chosen orangutans into the forest, and a corps of volunteers to carry them from the forest helipad to K6. And media coverage, of course.

In total nineteen orangutans were to be released. All of them had been transported to K6 early in May, but they were freed in two phases because of the size and diversity of the group. The idea was to make the competition less fierce by letting timid orangutans get established before the tougher ones came out. So thirteen orangutans were freed on May 16. Five tough guys got out a week later.

On the morning of May 16 a rope was pulled to release the cage lock, the door sprang open, and out climbed the orangutans. That was probably their last shared experience, because each took it from there in a different way. Some climbed off the cage into the foliage, others down the ladder to the ground. Within an hour the more knowledgeable, and

Inside the cage, Tono gets ready to leave.

Jaja, facing the strange new world of an endless forest, decided she'd had enough of the picnic and would just go home now. She started back along the path, walking upright.

perhaps the naively reckless, headed like pioneers into the forest. Several slipped off so quietly I didn't notice them. Mojo stood out—she was a whiz, racing to explore the forest and finding lots of things to eat. The braver orangutans followed the pioneers, our technicians trailed after them, and most of the remaining orangutans straggled along behind the technicians. It all seemed completely aimless. No one knew where to go or what to do.

Naive or not, the more adventurous orangutans dashed boisterously through the trees. Most sampled forest foods like fallen fruit, leaves, rattan shoots, and termites. About noon, the hottest part of the day, they all flopped down. Mojo bit off some palm leaves and made herself a ground nest. Boyke and Siti reclined together under a shady palm. At that moment, on their first day of freedom in a beautiful green forest, they looked supremely happy.

Not all was joyous. One orangutan came out of the cage, looked around, climbed back in, and shut the door. Sylvie balked at following the crowd into the forest, but once alone she howled in distress until a technician returned and coaxed her along. Jaja kept turning around and heading back down the trail—out of the forest—as if she'd enjoyed the picnic but was going home now.

Even those who were raucously enjoying themselves had little sense of how to travel in the forest. They had trouble using trees and lianas, probably because they had climbed on unbreakable chains at Wanariset. Every few minutes there was a loud crack as a tree broke and the innocent plummeted down with a crash. Siti boldly climbed on a large spiny rattan and got stabbed before she wised up and backed out. Shaulie and Ida sliced their hands on the sharp edges of bandang stalks. Many had trouble planning travel routes: they'd head off through the trees only to come to an impassable gap, with no option but to retrace their route or stoop to ground travel.

As for foods, they mostly tried easy ones like leaves, shoots, and small fruits. They also made mistakes. Boyke tried big coconuts—fine in

Mojo, one of the whizzes of the group despite her crooked right hand, seemed right at home in the forest and raced off to eat, climb, and explore.

principle but not in practice, because the shells are too hard. A couple tasted mushrooms and the wrong kind of soil; given the chance of poison, the mushrooms might have been a fatal mistake and the wrong soil could have made them ill. But even counting their errors, they tried very few foods. Even more worrying, some of them ignored perfect opportunities to learn about forest foods. Jaja simply sat and waited for the picnic to arrive. Shaulie clearly saw Mojo eat rattan shoots and should have rushed in to study or scrounge. Instead she walked right through Mojo's rattan as if it didn't exist.

Social relations were a similar jumble. Boyke, Tono, and Jaja immediately approached humans as if all that time living with orangutans at Wanariset counted for naught and life with humans was now back on the agenda. Boyke kept trying to play with humans (shades of his "owner's" emphasis on playing with him). Tono and Jaja kept sidling up to women, as if their most important humans had been females.

Joyful or distressed, none of these orangutans knew that their new situation was permanent. The days of human-provided food and shelter were over. Technicians did lead them back to K6 at the end of the day for a second ration of food, to show them that K6 was the place to come for support. That done, humans left for the day and the orangutans were on their own.

Settling in

We came back daily to provide provisions and monitor progress. Within a week new patterns started to emerge. Siti drew together with Kiki and Ida; the three naive little females revived the friendship they had established when they were cagemates two years earlier. Siti had lost her close companion from Wanariset, Boyke, because he broke a leg falling out of a tree a week after his release and was sent back for repairs. The three females probably stuck around K6 because they were barely past infancy

and clung to humans for parent-like protection, reassurance, and guidance. They also still depended heavily on provisions, because they could handle only very simple feeding and travel. Tono fast became notorious as the group bully and the only orangutan to gain weight (because of all the bananas he stole).

Then the tougher five were freed. As a result, several of the smaller orangutans began to avoid K6, probably out of fear, then vanished. The tough guys had their own adjustments to make. Judi, already about thirteen years old, hid up a tree. She was the only one in the group who seemed to realize that the move was serious. Petai, the oldest male, bit one of the technicians. Petai had been badly abused in captivity; our fears that he would be dangerous turned out to have been well founded.

Panjul fast became the dominant male. There was little dispute about it. Petai had dominated Panjul at Wanariset, but Panjul was far more adept in the forest. Petai simply retired to the north. Panjul wrestled all the other males down one-handed (they were mere juveniles), then rounded up as many of the females as he could. Jaja fled at the sight of him and Judi wouldn't give him the time of day, but Mojo took a shine to him and the two struck up a liaison that was to last over a year. As for the more naive orangutans, they followed Panjul's lead through the forest. Those that were not too timid, like Siti and Tono, pestered him to share his food and expertise.

After a month, Imelda and Aming arrived. They'd raided the village again, this time demolishing the villagers' electrical system as well as their fruit. For their transgressions, the two were banished to distant K6. Imelda vanished from K6 within a day. Aming stayed for weeks, even though Panjul plagued his first days. Panjul was a bit younger than Aming and amiable, but he was the dominant male. They were bound to meet, and when they did, it was sure to be tense.

The two saw each other at provisioning on Aming's third day. Panjul tried seven times that morning to approach Aming. He made every effort

Tono's interests fixed on Femke, a young student, rather than on the forest.

to be nice, and showed no signs of aggression, but Aming still backed off. Panjul did, gradually, get closer. Initially he could come no closer than thirty meters, but by noon he came close enough to touch.

Then, on the verge of making contact with Aming, Panjul abruptly turned aside and started eating leaves right before Aming's eyes. It made no sense, when he was at the point of achieving what he'd worked for all morning—meeting Aming! But he kept this up for another hour. Once he even turned his back, as if so engrossed in eating that he completely forgot Aming was there. Then came a tense moment when the two were only five meters apart. Aming suddenly reclined on his back, as if relaxed, and *he* started to eat too. What was even stranger, he ate something these orangutans never ate, the leafstalk of a small palm, *Licuala spinosa*. Panjul watched Aming eat, then after Aming retreated, Panjul ate from the very leafstalk Aming had eaten, and reclined on his back too. Just ten minutes later, Panjul approached Aming very, very, slowly, squeaking softly. This time, when Aming did not flee, Panjul touched his foot. Aming flinched but extended a hand to Panjul, and the two began to grapple cautiously.

As far as I could figure it, here were orangutans deliberately pretending to eat, pretending to ignore one another, and pretending to relax. If eating is the best activity orangutans can imagine and if ignoring a companion is the best social compliment they can pay, a sign perhaps of keeping the peace, then Panjul was signaling friendly intent. When Aming ate in front of Panjul, he seemed to be signaling his acceptance of Panjul's gesture, because it was just minutes later that the two finally made contact. The sequel to their meeting is a shade less touching. Panjul never thrashed Aming the way Charlie thrashed other males, but he made sure Aming knew who was boss. Every day for several weeks Panjul wrestled Aming. The veneer was of play, but Panjul always finished on top.

Aming was a great addition to the K6 community because his forest expertise shone through. He learned his way around amazingly quickly, independently finding other orangutans that were out of sight and navigating new areas of forest within two or three days of his arrival. His knowledge greatly helped the naive younger orangutans like Kiki, Ida, Tono, and Siti. In a matter of days they discovered Aming was a meek, mild-mannered food guru. If he found and prepared a difficult food, like a rattan heart or a coconut, these youngsters swarmed him. Several times the whole pack of them literally sat on his head until he gave up the food and left it to the lot of them.

Taking stock of forest skills

Things were settling down by the second month. Most of the orangutans we monitored knew how to obtain provisions and deal with humans, as well as how to find safe sleeping places, navigate the nearby forest, and forage.

The orangutans that we could monitor were gradually ranging farther from K6, hundreds of meters in all directions, along routes they made themselves. Often Panjul and or Mojo took the lead on excursions away from K6 and the know-nothings tagged along at the rear.

Other orangutans, however, had vanished and it could have been because their navigational skills were inadequate. Joshua, after disappearing from K6, showed up at a nearby village and ransacked a villager's home. An old trail near K6 led directly to the village Joshua hit; he had apparently lost his bearings and followed the trail instead. Jaja got lost, probably the same way, and suffered worse consequences. After following others far to the west one day, she didn't return. She usually followed technicians back to K6, but that evening she didn't, and then probably couldn't find her way back on her own. All she knew how to eat was rattan shoots, and so she almost starved. She finally surfaced ten days later west of the forest, emaciated, parasite-infested, and depressed. She survived because a kindly villager took her in and called Wanariset for help.

One day Ida forgot to follow her friends, found herself alone, and spent several forlorn hours with no idea what to do.

I saw Ida get lost. One morning she joined others who were eating fruit high in a tree. After half an hour they began to climb down and leave, one by one, and head north. Ida was the last one down. She looked for the others but by then they had disappeared. That gave her pause, but she started north anyway. I knew she'd picked the right direction but she seemed uncertain, moving slowly and pausing to look and listen. Once sounds came from the west, so she harkened and took a few steps that way. Then sounds came from the north, so she peered over there and moved a couple of steps that way, but again hesitated. Finally she came to me, her face very worried, made a little squeak, and put her hand on my knee. All I could do was pat her encouragingly and stay nearby. Unfortunately, Ida also stayed near me, so we were stuck to that spot for two hours. By mid-afternoon I took pity and made motions towards K6, as a suggestion about what she might do. She brightened, hopped up, and followed. I hoped she'd take the lead, but when I stopped, she stopped. In the end I had to lead her all the way back to K6.

Because of Enggong, Paul, and Bento, I knew many of the orangutan foods and how to obtain them. So with the newcomers, I could see who knew what about food. Some, like Panjul, Mojo, and Liar, knew what they were doing. They headed straight for forest foods within minutes of leaving the cage, even foods I knew to be difficult to find and hard to eat once they were found. Most of the others, however, knew little.

One big problem appeared to be recognizing forest foods, because the newcomers ate very few of the foods available. Judi ate ants and leaves but ignored the termites, bandang hearts, and even fruits that wise orangutans favored. Jaja didn't try any forest foods at all for the first few weeks. She finally learned to eat rattan shoots only because a woman student she liked, Femke den Haas, taught her. Femke pulled a rattan shoot, tore open its sheath at the base, and ate the tender material inside while Jaja watched. Jaja then took a rattan leaf and tore it apart, but ate the tip instead of the base. Femke showed her again how to do it correctly then gave Jaja the shoot she'd pulled. Jaja ate it immediately then pulled one on her own, and this time ate it correctly. She spent the whole rest of the day going from rattan to rattan, pulling shoots and eating them. Orangutans are unparalleled food addicts so not eating forest fare makes sense only if they don't recognize it as food.

A second problem was that the newcomers couldn't know where to find forest foods initially, even if they recognized the foods, because none of them had ever been in this forest before. Some key foods, like rattans, were everywhere, but other key foods, like fruits and some defended foods, were hidden or dispersed. For these, orangutans have to know their specific location. If Aming was any example, orangutans who come to a new area follow residents to learn where the foods are. K6 newcomers had tried that, but it didn't work initially because there were no knowledgeable resident orangutans around K6. So the newcomers had to explore on their own. Most explored within about a one-hundred-meter radius of K6 for the first few weeks, then gradually broadened their range as they gained knowledge. That effectively brought them knowledge of an increasingly broad range of places where various foods could be found. Once they had come across a good food source, some of them at least would remember its location and return to it. Mojo and Panjul reportedly knew of a good termite area to the north of K6 and returned there on a regular basis. Once somebody acquired the knowledge, others could learn from them. That meant knowledge about food places grew slowly in the beginning but would probably accelerate.

The final facet of the food problem, obtaining foods once they'd been identified and located, revealed the same range of skills in these newcomers—ignorance to expert. Among the experts, Mojo whizzed through dozens of different foods within her first two days of freedom, and Panjul efficiently extracted the heart of one of the largest rattans I have ever seen, within two hours of his release. These feats left me puzzled. I knew from K3 that it took years of experience for

orangutans to acquire the skills to obtain difficult foods, and these K6 newcomers had been confined in cages for years, what with captivity, then quarantine and socialization. The only explanation I could see was that they must have acquired the expertise in their early lives as wild orangutans, before they were captured. They were resurrecting their expertise from memory. Herman Rijksen and Helga Peters both reached the same conclusion.

The others faced problems. Some, like Ida and Siti, were small and weak. When they tried to pull bandang spears, they often failed and left empty-handed. Some concocted and clung to wrong-minded strategies. Tono tried for months to open termite nests by banging them against hard surfaces, although it never, ever, worked. He once tried Aming's skull as his hard surface. Human observers were highly amused but Aming was furious, and it still didn't work. Others used techniques that worked but were hopelessly inefficient. Judi ate ants by delicately picking them up one by one, between thumb and forefinger, and placing each carefully in her mouth. Sensible orangutans mopped up a whole row of ants with one swipe of the back of their hand. It looked as if it would be a long haul before they got past these problems: you can't speed growth or unstick pigheadedness.

We had certainly expected that most of the ex-captives would know little about forest life when first released, and that they faced a lot of learning. In many cases, they had to learn the dangerous way, by trial-and-error, because there were no experienced orangutans in their area for them to learn from. They made some great leaps forward that way, even within two months. Kiki, for instance, discovered all on her own how to chew palm leafstalks for juice.

Learning about new foods from others is generally a safer strategy than trial-and-error experimentation. Scrounging foods that others had found and prepared was a very important version of that strategy for orangutans, one that I'd observed them use around K3. Scrounging became an efficient way to build expertise around K6 too, once a few

Siti, like Tono, somehow got it into her head that termite nests could be opened by banging them, and both orangutans stubbornly persisted in this mistaken approach for months.

discoveries had been made. Siti and Ida, for instance, learned about palm leafstalks that way: once Kiki discovered how to eat them, these two scrounged bits and pieces by pestering her until she shared. That scrounging allowed them to taste this new food and offered clues about how to obtain it. After a couple of weeks of scrounging palm leafstalks from Kiki, Siti had mastered the basic procedure herself and had graduated to mature palms. On her first try at a mature palm, she devoted five hours to working one single leafstalk.

Siti, in fact, looked like the model forest pupil. She'd been blessed with the perfect personality—motivated, assertive, curious, energetic, good-humored. On top of that she had a systematic approach for learning. She paid keen attention to other orangutans' foraging and raced in to beg and scrounge when they found something. She was even bold enough to crowd in on Panjul and insist on having some of whatever he found; amazingly, he let her. This opened the door for her to share his wealth of knowledge. She practiced diligently, adding to and fixing up what she'd gleaned from others. Dead naive though she was at the outset, she had an approach that set her squarely on the fast track to success.

By the time I left for the year, the score looked like this: as with other releases, half the group had disappeared from K6 after one month, and only six remained after two (not counting Aming)—Ida, Siti, Judi, Mojo, Panjul, and Dan. The others had either returned to Wanariset because of health problems or they disappeared in the forest. Those who had to return to Wanariset were Boyke, with his broken leg, Tono with mysterious slash wounds, Kiki with sudden vomiting and blood in her feces, and Jaja with injuries she'd sustained while alone in the forest. Of those that had vanished, hopes were that they were living normal forest lives. Some, like Liar, probably were; others had likely met with mishaps. Shaulie and Sunni, for instance, had shown signs of knowing very little about forest life; survival, for them, was unlikely. The orangutans still at K6 were coping reasonably well. Whether their skills would support them independently through difficult times, like the irregular long dry seasons, was less certain. Mojo, Dan, and Panjul were all well versed in forest skills and might be able to succeed independently, at least under normal conditions. But Siti and Ida were still beginners in terms of expertise, even if they were well on their way. Judi, while hanging on, was progressing painfully slowly. It looked as though it could be a long, slow road to independent forest life.

THE MIRROR CRACK'D...

I arrived at Wanariset in 1997 to face the prospect of one of the long dry seasons that hit Borneo once a decade or so. It was only June, but the rains had already stopped. The probable culprit was El Niño, the ocean current from the west coast of South America that swells up every few years to wreak havoc over much of the world. In 1997, El Niño supposedly caused the moisture-rich winds that blow west over the Pacific to drop their rain sooner than usual, so rain that normally drenched Borneo fell into the ocean instead. The long dry spell was expected to bring forth a glorious massive fruiting. But before that it brought privations.

The dryness didn't seem so earth-shattering at first. It was a relief to be spared the mosquitoes, leeches, and mould; and indeed, much in the orangutans' lives remained normal. At K3, Enggong was still efficiently good-natured, Bento a bully, Paul sober-sided, Charlie belligerent. Patterns were changing, but that was as it should be, as they left behind childish things. Bento, Enggong, and Paul no longer tolerated one another. Bento had moved west, Enggong had shifted south, probably nearer his girlfriend, Tuti. Paul now reigned at K3, but only because nobody else wanted the place. Even Charlie's visits were rare. Everyone was in good health, except Bento. He'd

Judi mothers Siti by grooming her.

Mojo and Panjul are an affectionate pair.

developed festering wounds, so he'd been sent to Wanariset for treatment. I saw him there, looking like the product of a bad encounter with a body piercer—holes all over, one of them clear through his cheek. But he was on the mend, and would soon be fit to go home to his forest.

As expected, few orangutans remained around K6—just Panjul, Mojo, Siti, and Judi—but they made a harmonious group. Panjul remained the benign leader and Mojo his affectionate partner. The group's real center was Siti. Even bolder in the forest at six than she had been in socialization at three, she brazenly tithed the fruits of Panjul's labor and plunked herself before Judi to be groomed. As for Judi, she stayed pretty much alone, except for Siti. She paid little attention to Mojo and gave Panjul the cold shoulder. Nearly adult, she probably liked it that way.

Technicians could offer only partial stories about a few others. Kiki, Boyke, and Jaja had returned to K6, cured of their ailments, but had soon disappeared. I was dubious about their succeeding on their own. They knew little about forest life—which was why they had had to return to Wanariset, after all—so their quitting K6 cut them off from their main source of support. For unknown reasons, Tono and Ida had been sent to Meratus, a new release forest far away, after recovering from health problems. Aming and Imelda had yet again found one another, and their way

Panjul, Mojo and Siti all share termites.

179

back to the village. This time they too were consigned to Meratus. There was no word of anyone else.

I especially wanted to know how their foraging had progressed. The news was pretty good. Around K3, Paul had mastered bandang hearts and handled termites like a master chef. He and Enggong, both nearing adolescence, were moving to efficiency and speed in their eating—cut the fine points, collect and eat more. The K6 crew were also learning. Mojo now drank her termites and Siti had learned about many new forest foods. Siti remained ignorant of important foods like palm hearts and termites, but by using Panjul and Mojo as food scouts, she lived very successfully beyond her expertise. Once, when Panjul was out on a flimsy branch eating fruit, Siti sailed in for her take. He must have been pushing the branch to its limit, because with the added weight, it snapped and down they both crashed. Without a murmur of complaint, Panjul picked himself up and simply went back to work. Judi, who was still eating ants one at a time, probably took more than a few hints from Siti. In fact, Siti and Judi intertwined their skills. Both ate palm pith in the same bizarre way, time after time, together or alone, so they had either developed the technique together or one had learned it from the other

The feeling of harmony soon began to fade. Bento was discharged from Wanariset after forty-eight days and, having committed no misdemeanors, was returned to K3. He was hardly in top shape when he arrived, after the transport and the Valium he'd been given to keep him calm. His first move was on to K3 to rest. Sadder yet, he seemed to have no idea what to do. Instead of going his own way, he acted like a newly arrived stranger and followed Enggong all day, eating whatever and wherever Enggong did. Bento unquestionably knew his way about—he'd ranged that area for three years and must have known its every centimeter. What he no longer knew, though, was the day-to-day pulse of the forest, its fruiting, its leafing, its goings on. That intimate sense of how things were changing in the forest was probably what normally fueled his daily plans, from where to head to what and who to look for.

Bento was so bereft of ideas that he even followed us. On his third day he followed us back to Post Sinaga, where he sat woefully on the porch, then slept on the roof. The next day we coaxed him back to K3, but within two days he was at the post again and looking south, the way out of the forest. This time he wouldn't return to K3. He headed south. As we feared, that was the end. Within a day he was seen at the swamp. Within two, he was apprehended in the village trying to climb into a car. I suspect he was trying to hitch a lift back to Wanariset; ex-captives in Sabah have been seen waiting for buses, then climbing aboard when they stop. Bento, always a lazy mooch, could have seen his stay in Wanariset as a luxury vacation—full-time repose with food delivered into his hands several times daily. Bento did return to Wanariset, but in a cage instead of a car, and not for long. Within a week he was on his way to Meratus, where there was less scope for village sorties. My last sight of him was departing, scrunched in a transport cage, atop the four-wheel-drive.

The harmonious world at K6 crumbled too. First it was Mojo. There had been rumors that she wasn't well, but she seemed normal when I saw her, still with Panjul and characteristically busy. She wasn't the madcap whirlwind she'd been the year before, but I took her sobering as a change befitting a working forest orangutan. She did one odd thing, pulling foods to the ground instead of climbing to get them, then eating them lounged on her back. I interpreted this as a nifty gimmick she'd invented to compensate for her crippled right hand. The year before I'd watched

Siti and Judi carefully chose palms with a liana (or a branch) passing through their crowns. The liana made a convenient perch, and it was also part of their system for opening the stalk. They bit the stalk open at the base of the leaf fan. The stalk, broken, flopped over the liana. The liana then made a sort of hanger that supported the stalk while Judi and Siti tore it open below.

What looks like a golden morning among the trees, with the sun streaming down through the morning mists, is in fact caused by smoke from the burning forest.

her struggle for minutes on end to find a perch where she could wedge herself securely in place so she could free her good hand for working the food. Working on the ground neatly eliminated that difficulty.

I was wrong. Gaby Fredriksson, who was studying bears in the forest, sent word late in September that something was seriously wrong with Mojo. She was resting on the ground too often and too long. So when I went to K6 two days later with Elke Meyfarth, my newly arrived student, we immediately checked. We arrived near noon when the orangutans should have been working, playing, or lazing about. Mojo was tightly curled up on the ground, eyes glassy. Sugianto, the technician, said she hadn't eaten much; yesterday she'd been better, but today was bad again. We realized that we had to take her to Post Djamaludin for monitoring. Sugianto simply carried her down.

Mojo ate that evening, but refused food the next morning. We radioed Dolin and together decided to send her to the clinic. He made it to the Post by 9:30 a.m., even with the two-hour hike from the village, carried her out, and had her taxied to Wanariset the same day. The news from Wanariset a few days later was that they were treating her, but within the week she had died. The vets had identified her problem, a chronic parasitic infestation. It had damaged her intestinal system irreparably. Mojo had been infected for at least three months, possibly even up to a year.

With Mojo's death we realized Panjul had been absent for six weeks, although he usually popped up every few days at K6 or Post Djamaludin. The death of his regular companion raised concerns that he might have been stricken with parasites too. There was another danger, perhaps worse—the drought. He might have left in search of food. An orangutan was spotted at the west side of the forest around this time; we figured it could have been him.

By then, in late September, the drought was serious and fires were springing up. The worst problem was fires deliberately set by humans. In Borneo, farmers use fire as part of their slash-and-burn system of shifting

agriculture: fire clears, ash fertilizes. On a small scale, it can be managed effectively. On a large scale, it damages forests and farmlands so much that they cannot regenerate. Large-scale burning is now common in Indonesian Borneo, as masses of transmigrants from other islands struggle to farm its infertile soils, and the holders of commercial concessions for mining, logging, and plantations set fires as an expedient way of disposing of unwanted forest.

Fires in 1997 were so extensive that the whole of Borneo, the third largest island in the world, was blanketed in smoke for months. It is now recognized as one of the earth's worst environmental disasters. The smoke, euphemistically called "haze," reached the Philippines, Singapore, peninsular Malaysia, and Thailand. In Kucing, Sarawak, the haze reduced visibility to five meters and sent pollution indices over 800. By international standards, values over 100 are unhealthy. Authorities advised people to stay indoors, issued face masks, and closed schools. Corporations evacuated their foreign employees—when they could, that is. Flights were canceled for weeks, and when they weren't canceled they were delayed for hours. For two full months, whether I was in Wanariset, Balikpapan, or the forest, the sky was never blue and the sun was nothing but a sullen orange ball sulking through the smoke. The smell of smoke was everywhere, even deep in the forest at dawn.

Sungai Wain mostly escaped the fires. But even without fire, its life was failing from the drought. Sitting in the forest, I sometimes heard what I thought was the tapping of gentle rain on leaves. Instead, it was the sound of dry leaves falling, falling by the thousands, from the drought-stricken trees. Streams turned to beds of sand; normally pliant lianas snapped at my touch; undergrowth wilted and died. The forest floor, normally shaded and cool, was every day more exposed to sunlight. I could no longer find enough shade to sit.

The orangutans' food supply was dwindling. Without water, new plant growth withered as it emerged or was eaten faster than it could be replaced. By October, fruit was visible on the trees, but it was still two

Judi, with her poor foraging skills, is emaciated because she was incapable of finding enough to eat in the difficult drought times.

months from ripening and the deepening drought threatened to shrivel it to death. Even termites, usually everywhere, dried up. By October, Paul checked chunk after chunk of nest, piece after piece of dead wood, but found nothing. Some days, the orangutans could locate nothing to eat but a few rattan shoots, palm leafstalks, termites, and ants. Part of the problem could have been expertise. Late in October I saw Charlie and Tuti scare up fifteen different foods in two days, compared to the half dozen that others found. Their extra three years' forest experience gave an advantage that might spell the difference between life and death.

The importance of those skills became painfully clear. Judi, with her pitiful skills, started to eat lounged on her back in October. After Mojo, that signaled waning energies to me. Then Judi came to K6 one morning for her bananas, ate them, lazed about for a while—but never left to forage. Periodically she got up as if to leave, but simply moved four or five meters then lay down again. About noon she held a large leaf towards us as if begging for food. Agus, the technician, said she did this occasionally, so we took it as a bad habit from captivity and moved away to discourage her. He said she usually gave up quickly. But she begged again, again, and again. By her sixth request we were worried. We looked closer. We'd all realized Judi was thin, but now all her ribs were showing and her stomach hollow. She looked pretty much at the end of her rope.

Although regulations forbade food beyond morning bananas, this was clearly an emergency. Agus placed his leftover rice on the ground for Judi. Judi, who had hardly moved all day, *ran* to get it. When she finished that, she got my biscuits then our water. Normally she ignored water, but that day she drank 1.5 liters in two minutes. Fortunately, Judi's close call ended happily. With extra food, technicians got her through the worst of the drought and with her energy restored, she was right back to work foraging on her own. I'd feared we were in for weeks of begging once the extra food began to arrive. To Judi's credit, she never even tried.

Only Siti sailed through with no problems. Having a small body to feed and everyone's indulgence probably helped. She just bounced along, never losing her good humor and insatiable curiosity. She especially liked Elke's rattan project. Elke had plots around K6 where she labeled every rattan plant and checked them for orangutan damage. Siti reveled in those labels. We periodically found collections of them that she'd stolen and stashed away. Siti also "helped": When Elke found a rattan with an intact shoot, Siti ate it. Or she made us her audience. Once she hid inside a leafy shell she'd made from two fan-shaped palm leaves, then burst forth, a sort of hairy Botticelli Venus. Other times she sailed into the air from a small tree and crashed down on her leaves, finishing with hearty raspberry sounds, or showered herself in sparkles of sand.

Until mid-November Siti was the only bright light in the picture. Then some fruit started to ripen, rain began to fall, and the drought finally broke. Even timid Judi blossomed, braving the climb up treacherous lianas to the tops of forty meter trees for durian fruit. Panjul returned to Post Djamaludin after a three-month odyssey. He was in poor shape and meekly crawled into a cage to rest, but after two weeks' convalescence was raring to go again. The worst seemed over. Most of the orangutans I knew had pulled through.

Inferno

The reprieve was short-lived. By February 1998, the drought had returned. Apparently little had been learned from the devastation wrought by the haze just months before. When the drought resumed, people resumed setting fires. This time the fires threatened Wanariset itself, right up to the clinic and the caged orangutans. The cages were by then crammed with 180 orangutans, many already refugees from drought and fire. In the end, all that saved them was a concrete fence.

Fires found Sungai Wain too. Gaby spotted the first fire up in the northwest. Others soon sprang up to threaten the forest from all directions for two whole months. Gaby scoured the region to marshal a fire-fighting team of fifty people. By sheer determination, they held back the fires. They cut firebreaks a mere meter wide with forest knives, swept them clean of debris with home-made brooms, then patrolled them night and day. The moment their vigilance flagged, the fires licked to life again. Burning trees toppled over the breaks to become bridges for crawling flames.

Heat threw blazing birds' nests high over the trees; then cooler air over unburnt forest let them fall again and ignite debris below. Fires even burrowed underground into coal beds where, unattended, they could find everlasting life. Just for shelter and a meal, the firefighters sometimes had to walk ten kilometers in the dead of night. More often, I heard, they worked until they sank to the ground exhausted, and slept where they fell.

It finally ended in mid-April when the only thing that could really

Fire razed thousands of hectares of forest.

Panjul, waiting for an opportunity to hit the post kitchen, makes himself comfortable.

stop the fires arrived: rain. When Gaby and her team could finally pause and take stock, they had saved about one-third of the forest, a core of about 3,500 hectares. As for orangutans, a few had been seen during the fires—sometimes they'd come to watch the firefighters—but we had no easy way of finding out who had survived. Provisioning had been abandoned when everyone had to turn their energies to firefighting, so there was nothing to attract the orangutans. We searched the forest for them, from one end to another, for two more months. We found two. I did meet most of the orangutans I knew within my first weeks in the forest, but only because they found us.

Panjul had resumed his K6-to-Post Djamaludin circuit; he was visible and in fine form. Judi reappeared, just before I arrived, several kilometers outside the forest by a chicken farm. That she had even survived was a hair short of miraculous. They'd returned her to the forest where food was reasonably abundant, near a second post, and there we spotted her raiding noodles from the kitchen. She was skeletal again, scruffy-haired and nervous, looking like she needed every noodle she could pilfer.

Early in June, I was near K3 with my assistant Iyan and I got up my nerve to go look at it. Fires had razed the forest right up to K3, Gaby had said; nothing was left. We entered the wreckage as we climbed the last hill—a ramshackle firefighters' shelter, great gashes of firebreaks, and a mangled mess of charred wood that had been Bento and Enggong's lush playground. K3 itself was untouched except for its ladder, which Charlie had apparently dismantled in some fit of pique. The old sitting log and the well-worn paths were still there, and the chewed-up palms, Paul's termite field, the clutter of old orangutan nests. But it was all so forlorn.

I had to sit down. No more than two minutes later, who strolled up from the west but Paul, as if it were any regular morning on any regular day at regular banana time. Our meeting wasn't magic, but it felt that way. Paul must have been close enough to hear us approaching K3, and, figuring K3 plus humans equals bananas, he showed up. Typically, he had detected us, but we hadn't detected him. In fact, that day brought a second unexpected bonus. Agus radioed that Tuti was in the valley just southwest of K3.

On another search, Iyan and I found fresh orangutan food remains, the shredded leafstalks of a small palm, daun biru (*Licuala spinosa*). It had to be Tuti, the only orangutan who ate this stuff and mangled it this way. We followed the food traces and found several more newly massacred palms. Then the trail went cold. I didn't mind. I was enjoying exploring

Tuti's world, a lowland along a small river. Then suddenly something drew my eyes—I don't know what, I was aware of neither sound nor movement—and I saw an orangutan, five meters away on the ground (not too good), hair erect (worse), running right at us (catastrophic). What escaped my lips was "Oh, my God, it's Charlie." And with that we ran, as fast as we could stumble, in the other direction. It was just my luck that Charlie ate daun biru the same way Tuti did. In my brief look I did manage to see that he was healthy, huge, and pugnacious as ever.

We found only one more orangutan, whom we couldn't identify. The guess was Uce, who was likely to range in that area and who had the skills to survive tough times. We never met Siti, Enggong, or any of the others reintroduced to Sungai Wain. Some may just have left, as Judi had, perhaps following a wild orangutan pattern of migrating with the food. Wild orangutans probably respond to both the pull of good food and the push of scarcities. Ours couldn't have had an alluring destination in mind: they knew nothing of what lay beyond Sungai Wain and we knew that it offered less, not more. But they could well have been driven out of Sungai Wain by scarcities during the drought.

Of the orangutans we did locate, we were stunned to find all but Judi hale and hearty (and Judi, although thin, was even alive). They were downright fat after the worst privations on record. How they beat the heat is a mystery, of which I got only a glimpse.

Some had adjusted their ranges, especially those whose original range had been devastated. Paul was seen far west of K3 during the fires. We met Tuti and collided with Charlie near rivers, probably the last havens of water and food during the drought. They had also altered their foraging because of food scarcities, although fruit had been plentiful from December to March (that probably accounted for the plumpness). Several had added new foods. Paul, Panjul, and Tuti ate from serdang palms and Tuti and Charlie ate daun biru, both items the orangutans had not previously eaten although they are abundant in Sungai Wain forest. The likely explanation is that they were pushed to identify new foods when regular foods dwindled. Even Judi learned a thing or two. She no longer ate ants one by one; like the others, she now grabbed a nestful.

The way Panjul altered his foraging was so Machiavellian that it must be told. He worked the Post Djamaludin staff, doing battle with them to reach the post kitchen. Staff would brandish weapons at him—blowpipes, slingshots, and a thermos. Empty or not, Panjul feared them all. He would retreat, advance again, then finally concede loss and retreat to the forest. Staff would proclaim victory, resume their ping-pong game or whatever else had been interrupted, and drop their vigilance. And every time I watched, Panjul was back within fifteen minutes, having circled around out of sight. He was often at the kitchen door before staff even noticed. Panjul even worked the kitchen at midnight, 2 a.m., and 5 a.m., when inattention was assured. We humans showed embarrassing intellectual deficiencies in the face of this ploy. We tried the reliable tactics of brandishing blowpipes and slingshots (the thermos was, helpfully, locked in the kitchen) but were shocked to find they didn't work. The reason was, it was pitch dark and Panjul couldn't see them. Nobody thought to shine light on the weapons instead of on Panjul's face.

The good news from these changes, amusing or not, is that the orangutans were able to adjust enough to survive. Captive-reared orangutans *can* readapt to the rigors of forest life, and that's important to know in assessing the effectiveness of rehabilitation. The other good news is that the terrible destruction in Sungai Wain mobilized human support. Wanariset organized an integrated survey to assess the drought and fire damage, aiming to develop a better management plan for the forest. Wake-up calls of this rudeness and violence should never be necessary. But at least this call was heard.

IF NOT NOW, WHEN?

Sungai Wain's orangutans are a minute fraction of the total population of orangutans worldwide, but their plight parallels that of all orangutans today. This tiny community, like the whole species, is on the verge of collapse, and this damage has been suffered at human hands. The threat is not new—it can be traced back 40,000 years to when humans first reached the isle of Borneo. But it is now on the point of sending orangutans into extinction, as it has sent so many other species in the last century.

Many efforts have been made to change our menacing practices. The human threat to orangutans has been recognized for most of the twentieth century, and many landmark steps have been taken to counter it. Indonesia has prohibited killing, trading, and possessing orangutans since at least the 1930s. The IUCN (International Union for the Conservation of Nature) and CITES (Convention on International Trade in Endangered Species) negotiated international protection for orangutans because of the risk of their extinction in 1975. Extensive research on orangutan adaptation has helped conservation authorities to design protection programs with a better understanding of orangutans' needs. In the 1990s, awareness of orangutans' endangered state all over the world has increased; conferences, documentaries, magazines,

Some of the logs waiting here to be shipped downriver are as large across as a tall man.

Gold was discovered in the sands beneath Tanjung Puting's forests. Fortune-seekers flocked in, razed hundreds of hectares of forest to sift its sands for gold dust, and polluted the river with the mercury they used to extract the gold.

and books have helped to spread the word. Worldwide foundations have sprung up with the mission of supporting orangutan survival, and broad-based conservation agencies like WWF and WSPA are coming to the fore to offer support.

For all that, as of 1999 things are probably worse for orangutans than they have ever been. Wanariset's ORP, only seven years in operation, has already taken in almost four hundred orangutans. Among them now are wild refugees from drought, fire, or development projects. To compare, Galdikas's Camp Leakey reportedly handled 100 to 200 ex-captives over twenty-five years. The IUCN estimates that orangutan habitat has decreased by more than 80 percent just in the last twenty years. Estimates of the wild population have dropped to 7,000–12,000 in Sumatra and 10,000–15,000 in Borneo. Human communities this size barely qualify as large towns. The totals also represent many tiny communities that live isolated from one another, so the population is even more fragile than these meager numbers suggest.

What's wrong is that the two greatest human threats to orangutans, predation and habitat destruction, are outpacing efforts to combat them. Direct predation still flourishes despite prohibitory laws. People eat orangutan meat, kill orangutan "pests," and capture orangutans to sell, especially since the recent climatic, political, and economic hardships in Indonesia.

The human development that has savaged Sumatran and Bornean rainforests since the 1960s poses a monumental threat to orangutans. The areas most coveted by humans are often lowland forests, which are prime orangutan habitat, and human uses commonly destroy the forest. Commercial ventures—from timber, gold, coal or oil extraction to rice, rubber, or oil-palm plantations—destroy such large areas of forest that resident orangutans are deprived of even the minimum of resources they need to survive. Immigrating humans, from transmigration programs and abroad, also clear forest. While small human settlements do little damage, tens of thousands of immigrants do a lot. To make it worse, tropical forests regenerate poorly, if at all, on cleared land. Cleared land is fast invaded by alang-alang grass which crowds out most other growth and defies eradication. The product is a wasteland.

If some human uses don't destroy forest, they still disturb or degrade it. Timber companies log selectively and have argued that this is non-disruptive and even improves habitat for some species. But selective logging still disrupts the delicate, complex interdependencies that support all life in tropical rainforests. Even careful selective logging, extracting about 10 percent of large trees, has been estimated to leave behind a patchwork of isolated micro-habitats unsuitable for 60 percent of the original flora and fauna. While disturbance on a small scale may be tolerable, commercial levels are unquestionably disruptive. And while some species may tolerate or even benefit from habitat alteration, it is known that orangutans do not.

To make it all more difficult, protection agencies have sometimes undermined their own effectiveness. Conservation agencies have hesitated to support programs dedicated to orangutans, because of the philosophy that what should be saved is ecosystems, integrated communities of plants and animals, not individual species. Unfortunately, protecting ecosystems may not insulate species targeted as prey—as orangutans are. Stopping direct predation would require measures focused on orangutans themselves. Conservation agencies have also tended to fall in with "sustainable development" and "integrated use" philosophies that advocate continued human extraction of forest resources and shared use of forests between humans and wildlife—at "controlled" levels. When the resources to be extracted or shared are resources orangutans need, as they often are, the inevitable result is greater vulnerability for orangutans, not protection.

Rehabilitation's growth since its birth in the 1960s may look like a mark of success. Energies and money have poured in, largely because orphaned orangutans win immediate sympathy from almost all onlookers. In fact, rehabilitation is now highly contentious as a form of orangutan protection.

Extensive human immigration into Borneo has swelled the population, especially along the rivers.

One criticism is that rehabilitation is simply not effective. Its original intent was to help law enforcement agencies stem trade in live orangutans. This goal has in the past appeared to be within reach, even to have been met, but confiscations are now increasing. International restrictions were thought to have slowed the export market, but incidents like those involving the Bangkok Six and Taiwan Ten, only seven or eight years ago, suggest that smuggling may simply have become less visible. It

has proven difficult to obtain stiff penalties then to have them upheld. Confiscation is often the only punishment meted out. As a deterrent, it has little impact.

The broader role of rehabilitation, returning ex-captives to free forest life, has been discredited as well. Few programs can provide credible success statistics and what information is casually available suggests success is far from assured. There is still little agreement on how to program rehabilitation. Some programs treat the risk of disease as paramount while others ignore it; some focus on social readaptation, others emphasize learning forest skills. Without monitoring, of course, it is impossible to weigh the various claims, and effective monitoring has proven exceptionally difficult.

A third arm of orangutan protection is education. Everyone recognizes that education about the importance of conserving natural areas and wildlife is more effective in the long run than policing can ever be. But in developing countries like Indonesia, where few know or care about the need for environmental protection, education has barely made a dent. Many Indonesians live close to the poverty line; all they think about is surviving from day to day. They also see foreigners and their own leaders made well-to-do by the very activities they are asked to renounce. The long-term advantages of conservation to the world at large are unlikely to carry much weight with them. As for educating the policy-makers, field workers, and business-political-commercial interests who have the most immediate and direct impact on orangutans, efforts appear to have been less active, even though it is in this area that the greatest benefits might accrue.

Protecting orangutans is becoming a harder job, not an easier one, and the question has certainly arisen whether it warrants the effort. Orangutans' precarious survival status, their apparent sensitivity, the large amounts of valuable habitat they need, and the costs of protecting them all weigh against protection.

There are equally weighty arguments in favor of protection. To be self-centered about it, orangutans are among our cousins, our closest living relatives. Many consider this close relation alone enough to favor protection. Not only that, only four species of non-human great apes remain alive on the earth, so orangutans offer one of the very few avenues in existence for understanding our own origins. If we eliminate them, we close an important window on ourselves.

In the broader realm of biological conservation, the orangutan is an important symbol. Many sites valuable to orangutans are valuable to other species too, so their protection assures support for other facets of nature as well. Orangutans also play a significant biological role. Many tropical rainforest plants need the help of animals to reproduce, because it is animals that disperse their seeds. Orangutans could be among the most important seed dispersers in Southeast Asia's rainforest ecosystems. They consume huge volumes of an exceptionally broad range of fruits, they can carry seeds great distances, and seeds passed through their digestive systems may germinate more effectively. Orangutans may also shape their forest because they break so much foliage. Among other things, their damage creates small gaps in the forest. Such gaps allow some plant species to grow; indeed, they are considered one of the main mechanisms by which tropical rainforests regenerate.

Beyond the question of whether we should protect orangutans lies another huge question. Can we? Orangutans are so perilously close to extinction that if their numbers fall much lower they will enter a spiral into inevitable disappearance. WWF-Indonesia recently estimated they could be extinct within a hundred years if we don't act to prevent forest fires and halt the destruction of Borneo and Sumatra's rainforests. WWF estimated that the 1997–98 drought and fires destroyed 40 percent of the orangutan habitat that remains; of that, millions of hectares are still slated to be cleared for plantations. Conservationists are racing to address all these challenges. Law enforcement is stiffening. Protection for some orangutan habitat, like Gunung Leuser Park in Sumatra, has recently become much more stringent. Rehabilitation efforts are growing: Wanariset is instituting forest-based halfway houses to ease ex-captives' transition to free lives, and a new program is being opened in central Kalimantan. The key is whether these measures are enough and whether they will take effect in time.

To that question, my best answer is one borrowed from David Quammen, a writer who has taken up the cause of many of the species vanishing from our planet. There are no hopeless causes—only hopeless people and expensive causes. Orangutan survival may be an expensive cause, but only hopeless people would abandon it.

If this seems a sad end to my tale, remember that the end of my tale is not the end of orangutans. Even as I dotted my last "i" and crossed my last "t," a tiny mote of hope arrived. Gaby sent word that Uce had appeared late in January 1999 with a baby, a healthy male, only about a month old, so conceived just as fires retreated. That tiny bit of orange fluff means many things—the first infant born in Sungai Wain, life reborn out of the ashes, and a future. Just as one swallow doesn't make a summer, one baby orangutan doesn't herald Eden. But if orangutans like Uce, whom we've dealt every bad card in the deck, can come through with such flying colors, surely orangutans deserve our respect and support, not condemnation and abuse. If we can find that respect, there is yet a chance. The real end of the tale is, after all, up to us.

ORGANIZATIONS CONCERNED WITH ORANGUTAN WELFARE

Balikpapan Orangutan Survival Foundation
  P. O. Box 447,
  Balikpapan 76103, Kalimantan Timur, Indonesia
  tel: +62 (542) 413069
  fax: +62 (542) 410365
  e-mail: bosbpn@indo.net.idwww.redcube.nl/bos/

Balikpapan Orangutan Society–U.S.A.
  BOS-USA
  P.O. Box 2113
  Aptos CA 95001-2113, USA
  www.orangutan.com

CITES (Convention on International Trade in Endangered
  Species)CITES Secretariat, 15, chemin des Anémones, CH-1219
  Châtelaine-Genève, Suisse.
  tel: (+4122) 979 9139/40,
  fax: (+4122) 797 3417
  e-mail: cites@unep.ch
  ww.wcmc.org.uk/CITES/english/index.html

International Primate Protection League
  P. O. Box 766,
  Summerville SC 29484, USA
  e-mail: ippl@awod.com
  www.ippl.org

IUCN (International Union for the Conservation of Nature)
  IUCN–CANADA OFFICE
  380 St-Antoine St. West, suite 3200
  Montréal PQ H2Y 3X7, Canada
  tel: (514) 287-9704
  fax: (514) 287-9057
  e-mail: poste@iucn.ca
  www.iucn.org

The Great Ape Project
  The Great Ape Project–International
  P. O. Box 19492,
  Portland OR 97280-0492, USA
  arrs.envirolink.org/gap
  Great Apes Canada
  849 East 4th St.,
    North Vancouver BC V7L 1K3, Canada
    tel: +1 (604) 983-3661
    fax: +1 (604) 983-8189

WSPA (World Society for the Protection of Animals)
  Headquarters,
  2 Langley Lane,
  London SW8 1TJ, UK
  tel: +44 (171) 793 0540
  fax: +44 (171) 793 0208

e-mail: wspa@wspa.org.uk

www.wspa.org.uk

WSPA Canada

44 Victoria St., Suite 1310

Toronto ON M5C 1Y2, Canada

tel: +1 (416) 369-0044

fax: +1 (416) 369-0147

email: wspacanada@compuserve.com

World Wildlife Fund for Nature

WWF International

Avenue du Mont-Blanc

CH-1196, Gland, Switzerland

tel: +41 22 364 91 11

www.worldwildlife.org

Zoocheck Canada Inc.

3266 Yonge Street, Suite 1729

Toronto ON M4N 3P6, Canada

tel: +1 (416) 285-1744

fax: +1 (416) 285-4670/696-0370

e-mail: zoocheck@idirect.com

www.zoocheck.com

IBLIOGRAPHY

Aristotle's Rubicon

Corbey, R. & Theunissen, B. (Eds.). (1995). *Ape, Man, Apeman: Changing Views since 1600*. Leiden: Department of Prehistory, Leiden University.

French, R. (1994). *Ancient Natural History*. London: Routledge.

Harrisson, B. (1987). *Orang-utan*. Oxford: Oxford University Press.

Janson, H. W. (1952). *Apes and Ape Lore in the Middle Ages and the Renaissance*. Vol. 20 of Studies of the Warburg Institute. London: Warburg Institute.

Salisbury, J. E. (1994). *The Beast Within: Animals in the Middle Ages*. New York: Routledge.

Yerkes, R.M. (1916). The mental life of monkeys and apes. *Behavior Monographs*, Vol. 3. New York: Holt.

Orangutans 101

de Boer, L. E. M. (Ed.), *The Orang Utan: Its Biology and Conservation*. The Hague: Dr. W. Junk Publishers.

Chevalier-Skolnikoff, S., Galdikas, B. M. F., & Skolnikoff, A. Z. (1982). The adaptive significance of higher intelligence in wild orang-utans: A preliminary report. *Journal of Human Evolution*, 11, 639–652.

Courtenay, J., Groves, C., & Andrews, P. (1988). Inter- or intra-island variation? An assessment of difference between Bornean and Sumatran orangutans. In J. H. Schwartz (Ed.), *Orang-Utan Biology*, pp. 19–29. New York: Oxford University Press.

Galdikas, B. M. F. (1979). Orangutan adaptation at Tanjung Puting Reserve: Mating and ecology. In D. A. Hamburg & E. R. McCown (Eds.), *The Great Apes*, pp. 194–233. San Francisco: Benjamin/Cummings.

Kaplan, G. & Rogers, L. (1994). *Orang-Utans of Borneo*. Sydney: University of New England Press.

Lethmate, J. (1982). Tool-using skills of orang-utans. *Journal of Human Evolution*, 11, 49–64.

MacKinnon, J. R. (1974). *In Search of the Red Ape*. New York: Holt, Rinehart & Winston.

Maple, T. (1980). *Orang-Utan Behavior*. New York: Van Nostrand Reinhold.

Miles, H. L. (1986). How can I tell a lie? Apes, language, and the problem of deception. In R. W. Mitchell & N. S. Thompson (Eds.), *Deception: Perspectives on Human and Nonhuman Deceit*, pp. 245–266. Albany, NY: State University of New York Press.

Rijksen, H. D. (1978). *A field study on Sumatran orang utans (Pongo pygmaeus abelii Lesson 1827): Ecology, behaviour, and conservation*. Mededelingen, Landbouwhogeschool Wageningen, the Netherlands: H. Veenan and Zonen B. V.

Rodman, P. S. & Mitani, J. C. (1987). Orangutans: Sexual dimorphism in a solitary species. In B. B. Smuts, D. L. Cheney, R. M. Seyfarth, R. W. Wrangham, & T. T. Struhsaker (Eds.), *Primate Societies*, pp. 146–154. Chicago: University of Chicago Press.

Schaller, G. B. (1961). The orang-utan in Sarawak. *Zoologica*, 46, 73–82.

Schwartz, J. H. (Ed.). (1988). *Orang-utan Biology*. New York: Oxford University Press.

Utami, S. S. & van Hooff, J. A. R. A. M. (1997). Meat-eating by adult female Sumatran orangutans (*Pongo pygmaeus abelii*). *American Journal of Primatology*, 43, 159–165.

van Schaik, C. P. & van Hooff, J. A. R. A. M. (1996). Toward an understanding of the orangutan's social system. In W. C. McGrew, L. F. Marchant, & T. Nishida (Eds.), *Great Ape Societies*, pp. 3–15. Cambridge, UK: Cambridge University Press.

The World According to Camp Leakey

Galdikas, B. M. F. (1995). *Reflections of Eden: My years with the orangutans of Borneo*. Boston: Little, Brown.

Galdikas-Brindamour, B. M. F. (1975). Orangutans, Indonesia's "People of the Forest". *National Geographic, 148* (4), 444–473.

Montgomery, S. (1991). *Walking with the Great Apes: Jane Goodall, Dian Fossey, Biruté Galdikas*. Boston: Houghton Mifflin.

The Sorcerer's Red Apprentice

Byrne, R. W. (1995). *The Thinking Ape*. Oxford: Oxford University Press.

Galdikas, B. M. F. (1982). Orangutan tool use at Tanjung Puting Reserve, Central Indonesian Borneo (Kalimantan Tengah). *Journal of Human Evolution, 10*, 19–33.

Parker, S. T., Mitchell, R. W., & Miles, H. L. (Eds.) (in press). *The Mentalities of Gorillas and Orangutans*. Cambridge, UK: Cambridge University Press.

Russon, A. E., Bard, K. A., & Parker, S. T. (Eds.) (1996). *Reaching into Thought: The Minds of the Great Apes*. Cambridge, UK: Cambridge University Press.

Russon, A. E. & Galdikas, B. M. F. (1993). Imitation in free-ranging rehabilitant orangutans (*Pongo pygmaeus*). *Journal of Comparative Psychology, 107*(2), 147–61.

Trouble in Eden

Freemon, S. (1994, January 4). Monkey business in Borneo. *The Guardian*, 14.

MacKinnon, J. (1977). The future of orang-utans. *New Scientist*, 74, 697–699.

Rijksen, H. D. (1982). How to save the mysterious "man of the forest"? In L. E. M. de Boer (Ed.), *The Orang Utan. Its Biology and Conservation*, pp. 317–341. The Hague: Dr. W. Junk Publishers.

Russell, C. (1995). The social construction of orangutans: An ecotourist experience. *Society and Animals*, 3(2), 151–170.

McGreal, S. (1993). "The Bangkok Six" —1990-present. *Ippl News*, 29(3), 21–22.

Spalding, L. (1997). *The Follow*. Toronto: Key Porter Books.

Wanariset: The New Rehabilitation

National Research Council (1998): *The Psychological Well-Being of Nonhuman Primates*. Washington, DC: National Academy Press.

Smits, W. T. M., Heriyanto, & Ramono, W. S. (1995). A new method for rehabilitation of orangutans in Indonesia: A first overview. In R. Nadler, B. F. M. Galdikas, L. K. Sheehan, & N. Rosen (Eds.), *The Neglected Ape*, pp. 61–68. New York: Plenum Press.

Yeager, C. P. (1997). Orangutan rehabilitation in Tanjung Puting National Park, Indonesia. *Conservation Biology*, 11(3), 802–805.

Zdziarski, J. M. (1997). Zoonotic diseases. In C. Sodaro (Ed.), *Orangutan Species Survival Plan Husbandry Manual*, pp. 108–112. Chicago: Chicago Zoological Park.

Tropical Rain Forest Homes

Byrne, R. W. & Byrne, J. M. E. (1993). Complex leaf-gathering skills of mountain gorillas (*Gorilla g. beringei*): Variability and standardization. *American Journal of Primatology*, 31, 241–261.

Etkin, N. L. (Ed.). (1994). *Eating on the Wild Side*. Tucson: University of Arizona Press.

Forsyth, A. & Miyata, K. (1984). *Tropical Nature*. New York: Scribner.

Jacobs, M. (1987). *The Tropical Rainforest*. Berlin: Springer.

Johns, T. (1990). *With Bitter Herbs They Shall Eat It: Chemical Ecology and the Origins of Human Diet and Medicine*. Tucson: University of Arizona Press.

Leighton, M. & Leighton, D. R. (1983). Vertebrate responses to fruiting seasonality within a Bornean rainforest. In S. L. Sutton, T. C. Whitmore, & A. C. Chadwick (Eds.), *Tropical Rain Forest Ecology and Management*, pp. 181–196. Oxford: Blackwell.

Matsuzawa, T. (1994). Field experiments on use of stone tools by chimpanzees in the wild. In R. W. Wrangham, W. C. McGrew, F. B. M. de Waal, & P. G. Heltne (Eds.), *Chimpanzee Cultures*, pp. 351–370. Cambridge, Mass.: Harvard University Press.

Povinelli, D. J. & Cant, J. G. H. (1995). Arboreal clambering and the evolution of self-conception. *Quarterly Review of Biology, 70*(4), 393–421.

van Schaik, C. P. & Fox, E. A. (1996). Manufacture and use of tools in wild Sumatran orangutans. *Naturwissenschaften, 83*, 186–188.

Whitmore, T. C. (1990). *An Introduction to Tropical Rain Forests*. Oxford: Clarendon.

Lord of the Flies

Byrne, R. W. & Whiten, A. (Eds.) (1988). *Machiavellian Intelligence: Social Expertise and the Evolution of Intellect in Monkeys, Apes and Humans*. Oxford: Clarendon Press.

Pereira, M. E. & Fairbanks, L. A. (Eds.). (1993). *Juvenile Primates: Life History, Development, and Behavior*. New York: Oxford University Press.

Russon, A. E. (1997). Exploiting the expertise of others. In A. Whiten & R. W. Byrne (Eds.), *Machiavellian Intelligence II: Extensions and Evaluations*, pp. 174–206. Cambridge, UK: Cambridge University Press.

Russon, A. E. (1997). The nature and evolution of intelligence in orangutans (*Pongo pygmaeus*). *Primates, 39*(4), 485–504.

Release Me

Clemmens, J. R. & Bucholz, R. (Eds.) (1997). *Behavioral Approaches to Conservation in the Wild*. Cambridge, UK: Cambridge University Press.

Kleiman, D. G., Beck, B. B., Dietz, J. M., & Dietz, L. A. (1991). Costs of a re-introduction and criteria for success: Accounting and accountability in the golden lion tamarin conservation program. *Symposium of the Zoological Society of London, 62,* 125–142.

Peters, H. (1995). *Orangutan Reintroduction? Development, use and evaluation of a new method: Reintroduction.* Unpub. M.Sc. Thesis, University of Groningen, the Netherlands.

The Mirror Crack'd…

Aglionby, J. (1998, December 27). Orangutans face extinction. *Guardian Weekly.*

Kaiser, R. G. (1997, September). Forests of Borneo going up in smoke. *Guardian Weekly.*

Choong, T. S. (1997, October 10). The reckless torching of Indonesia's forest lands. *Asiaweek.*

Economy and Environment Programme for South East Asia (EPSEA) and World Wildlife Fund (WWF). (1998, June 4). The Indonesian fires and haze of 1997: The economic toll. Available: www.idrc.org.sg/eepsea/fire.htm [1998, June 4].

Integrated Forest Fire Management (1998, April 21). Causes of forest fires. Available: http://smd.meta.net.id/iffm/background.html [1998, April 21].

Leighton, M. & Wirawan, N. (1986). Catastrophic drought and fire in Borneo Tropical Rain Forest associated with the 1982–1983 El Niño southern oscillation effect. In G. T. Prance (Ed.), *Tropical Rain Forests and the World Atmosphere* (AAAS Selected Symposium 101), pp. 75–102. Boulder, CO: Westview Press.

If Not Now, When?

Aveling, R. J. & Mitchell, A. H. (1982). Is rehabilitating orang utans worth while? *Oryx, 16*(3), 263–271.

Benirschke, K. (Ed.). (1986). *Primates: The Road to Self-Sustaining Populations.* New York: Springer-Verlag.

Cleary, M. & Eaton, P. (1992). *Borneo: Change and Development*. Singapore: Oxford University Press.

Frasier, J. G. (1997). Sustainable development: Modern elixir or sack dress? *Environmental Conservation, 24*(2), 182–193.

Galdikas, B. M. F. (1982). Orang utans as seed dispersers at Tanjung Puting, Central Kalimantan: Implications for conservation. In L. E. M. de Boer (Ed.), *The Orang Utan. Its Biology and Conservation*, pp. 285–298. The Hague: Dr. W. Junk Publishers.

IRIP News Service (1998, November 23). Politics and peat: The one million hectare sawah project. Available: www.serve.com/inside/edit48/1juta.htm [1998, November 23].

Nadler, R., Galdikas, B. F. M., Sheehan, L. K. & Rosen, N. (Eds.) (1995). *The Neglected Ape*. New York: Plenum Press.

Peterson, D. (1989). *The Deluge and the Ark: A Journey into Primate Worlds*. Boston: Houghton Mifflin.

Primack, R. & Lovejoy, T. E. (Eds.) (1995). *Ecology, Conservation, and Management of Southeast Asian Rainforests*. New Haven: Yale University Press.

Quammen, D. (1996). *The Song of the Dodo: island biogeography in an age of extinctions*. New York: Scribner.

Wilson, W. L. & Wilson, C. C. (1975). The influence of selective logging on primates and some other animals in East Kalimantan. *Folia Primatologica, 23*, 245–274.

INDEX